ENVIRONMENTAL CONSCIOUSNESS, NATURE AND THE PHILOSOPHY OF EDUCATION

This book explores alternative ways of understanding our environmental situation by challenging the Western view of nature as purely a resource for humans.

Environmental Consciousness, Nature and the Philosophy of Education asserts that we need to retrieve a thinking that expresses a different relationship with nature: one that celebrates nature's otherness and is attuned to its intrinsic integrity, agency, normativity, and worth. Through such receptivity to nature's address, we can develop a sense of our own being-in-nature that provides a positive orientation towards the problems we now face. Michael Bonnett argues that this reframing and rethinking of our place in nature has fundamental implications for education as a whole, questioning the idea of human "stewardship" of nature and developing the idea of moral education in a world of alterity and non-rational agents.

Drawing on and revising work published by the author over the last 15 years, this book will be essential reading for students and scholars of environmental studies, environmental education, and the philosophy of education.

Michael Bonnett has published widely in the field of philosophy of education, giving particular attention to ideas of learning, thinking, personal authenticity, and the character of the teacher–pupil relationship in education. His book *Children's Thinking: Promoting Understanding in the Primary School* (1994) explored the importance of poetic thinking for education. More recently, his focus has been on aspects of sustainability and environmental education, including developing a phenomenology of nature and exploring ways in which human consciousness is inherently environmental. His book *Retrieving Nature: Education for a Post-Humanist Age* was published in 2004, and his edited collection *Moral Education and Environmental Concern* was published in 2014 by Routledge.

Research and Teaching in Environmental Studies

This series brings together international educators and researchers working from a variety of perspectives to explore and present best practice for research and teaching in environmental studies.

Given the urgency of environmental problems, our approach to the research and teaching of environmental studies is crucial. Reflecting on examples of success and failure within the field, this collection showcases authors from a diverse range of environmental disciplines including climate change, environmental communication and sustainable development. Lessons learned from interdisciplinary and transdisciplinary research are presented, as well as teaching and classroom methodology for specific countries and disciplines.

Titles in this series include:

Sustainable Energy Education in the Arctic
The Role of Higher Education
Gisele Arruda

Interdisciplinary and Transdisciplinary Failures
Lessons Learned from Cautionary Tales
Edited by Dena Fam and Michael O'Rourke

Environmental Consciousness, Nature and the Philosophy of Education
Ecologizing Education
Michael Bonnett

For more information about this series, please visit: www.routledge.com/Research-and-Teaching-in-Environmental-Studies/book-series/RTES

ENVIRONMENTAL CONSCIOUSNESS, NATURE AND THE PHILOSOPHY OF EDUCATION

Ecologizing Education

Michael Bonnett

Routledge
Taylor & Francis Group
LONDON AND NEW YORK

earthscan
from Routledge

First published 2021
by Routledge
2 Park Square, Milton Park, Abingdon, Oxon OX14 4RN

and by Routledge
52 Vanderbilt Avenue, New York, NY 10017

Routledge is an imprint of the Taylor & Francis Group, an informa business

British Library Cataloguing-in-Publication Data
A catalogue record for this book is available from the British Library

Library of Congress Cataloging-in-Publication Data
A catalog record for this book has been requested

ISBN: 978-0-367-37343-6 (hbk)
ISBN: 978-0-367-37344-3 (pbk)
ISBN: 978-0-429-35322-2 (ebk)

Typeset in Bembo
by Apex CoVantage, LLC

CONTENTS

PREFACE

A central theme developed in this book is that our efforts to address the environmental crisis that now besets us frequently have been stymied by the kind of thinking that has become dominant in Western culture and policymaking. This thinking essentially regards nature purely as a *resource*. It is argued that we need to understand the philosophical sources and fuller character of this kind of orientation in order to clear a space for a different relationship with nature: one that will enable us to approach environmental issues in a way that is better informed and more truthful. Elucidating the character of this way of experiencing nature promises to reveal extensive implications for how we think of human flourishing and education. In many respects, this endeavour represents a continuation of a project that I began in a previous book (*Retrieving Nature: Education for a Post-humanist Age*, 2004) and some of the arguments presented in the current work build on ideas that were set out in that book. Where this is the case, I give sufficient explanation for the current account to be "stand-alone". The current work takes the opportunity to bring into one narrative – and to extend – ideas that I have presented in a variety of research journals and edited collections over the last 15 years or so.

With regard to the way that arguments are developed, it might be helpful to mention that their structure is "spiral" rather than purely linear. This is to say that rather than being set out in a series of completed steps, key ideas frequently are revisited and further elaborated as the narrative proceeds, responding to the invitations and demands of evolving contexts. While it would be incorrect to say that this was inspired by the advocacy of a "spiral curriculum" by the influential educational psychologist Jerome Bruner in his classic *The Process of Education* (1963), it is certainly consistent with this idea. Also, it occurs to me that this form of argument reflects better the way that nature reveals itself, and the way in which we learn most from it. Things in the natural world do not show themselves all at once. On any one viewing the countenances that they present hide others.

Our understanding of nature grows not by attempting to derive it from one experience and to encapsulate this in one description that is taken to be definitional, but rather by revisiting the phenomenon under consideration on numerous occasions and varying circumstances and contexts, and being prepared to hold a multidimensional perspective. In this way, we can develop a growing feel for the myriad and complex interrelationships in which things in nature are embedded. Indeed, it is argued that in an important sense, it is these relationships that constitute things in nature. Finally, while some of the philosophical content is demanding, my intention has been to ensure that ideas and any special terminology are sufficiently explained and illustrated so as to produce an account that is accessible to those without specialist knowledge of environmental philosophy.

ACKNOWLEDGEMENTS

Parts of this book draw on work that I have previously published and I am grateful for publishers' permissions to use this material. The articles from which material is drawn and the chapters to which this material contributes are as follows:

Chapter 1

Bonnett, M. (2013) Sustainable development, environmental education, and the significance of being in place, *Curriculum Journal*, 24(2), pp. 250–271. Copyright 2013 British Curriculum Foundation, permission given by Taylor & Francis www.tandfonline.com on behalf of British Curriculum Foundation

Chapter 2

Bonnett, M. (2007) Environmental education and the issue of nature, *Journal of Curriculum Studies*, 39(6), pp. 707–721. Taylor & Francis www.tandfonline.com.

Chapter 3

Bonnett, M. (2013) Normalizing catastrophe: Sustainability and scientism, *Environmental Education Research*, 19(2), pp. 187–197. Taylor & Francis www.tandfonline.com.

Bonnett, M. (2009) Systemic wisdom, the 'selving' of nature, and knowledge transformation: Education for the 'greater whole', *Studies in Philosophy and Education*, 28, pp. 39–49. Springer Nature.

Chapter 4

Bonnett, M. (2017) Environmental Consciousness, sustainability, and the character of philosophy of education, *Studies in Philosophy and Education*, 36(3), pp. 333–347. Springer Nature.

Bonnett, M. (2009) Schools as places of unselving: An educational pathology? In G. Dall'Alba (ed.) *Exploring Education Through Phenomenology: Diverse Approaches* (Chichester, John Wiley & Sons).

Bonnett, M. (2013) Normalizing catastrophe: Sustainability and scientism, *Environmental Education Research*, 19(2), pp. 187–197. Taylor & Francis www.tandfonline.com.

Chapter 6

Bonnett, M. (2015) Transcendental nature and the character of truth and knowledge in education, in P. Kemp and S. Frolund (eds.) *Nature in Education* (Zurich, LIT-Verlag).

Bonnett, M. (2009) Systemic wisdom, the 'selving' of nature, and knowledge transformation: Education for the 'greater whole', *Studies in Philosophy and Education*, 28, pp. 39–49. Springer Nature.

Bonnett, M. (2013) Self, environment, and education: normative arisings, in M. Brody, J. Dillon and R. Stevenson (eds.) *International Handbook of Research on Environmental Education* (New York, Routledge). Reproduced by permission of Taylor and Francis Group, LLC, a division of Informa plc.

Chapter 7

Bonnett, M. (2007) Environmental education and the issue of nature, *Journal of Curriculum Studies*, 39(6), pp. 707–721. Taylor & Francis www.tandfonline.com.

Bonnett, M. (2012) Environmental concern, moral education, and our place in nature, *Journal of Moral Education*, 41(3), pp. 285–300. Copyright 2012 Journal of Moral Education Ltd, permission given by Taylor & Francis www.tandfonline.com on behalf of Journal of Moral Education Ltd.

1

NORMALIZING CATASTROPHE

A backdrop to environmental issues

The purpose of this chapter is to set out in a preliminary way a number of influential motives, ideas, and views that shape our current understanding of, and responses to, our environmental situation in the West. The position taken in this book is that this understanding and key ideas that have emerged such as sustainable development are conditioned by a range of environing concepts and social/cultural/political concerns and attitudes that require philosophical examination as an aid to discerning a defensible way forward in addressing our environmental situation. Given the complexity and variety of factors in play, the selection of ideas presented in the following sections is necessarily partial, but it does, I believe, constitute a powerful constellation of thought that is salient in environmental debate.

A new epoch?

As we make our way deeper into the twenty-first century, we find ourselves confronted by environmental problems of unprecedented magnitude – and this, notwithstanding the fact that many of these problems have long been recognized and that there have been ongoing attempts to address them. Environmental pollution, habitat destruction, species extinction, resource depletion, and climate change are now only too familiar examples. They occur on a massive scale and for the most part continue unabated. This is not to say that there have not been some successes in addressing environmental problems, but they have been limited, and overall the situation remains very troubling. For example, the United Nations Environmental Programme (UNEP) *Global Environmental Outlook Report*, GEO 5, of 2012 highlights expanding dangers of anthropogenic impacts:

> As human pressures within the Earth System increase, several critical thresholds are approaching or have been exceeded, beyond which abrupt and

> non-linear changes to the life-support function of the planet could occur. This has significant implications for human well-being now and in the future. For example: climate variability and extreme weather influence food security; crossing of thresholds leads to significant health impacts, as shown by the increase in malaria in response to rising temperatures; increased frequency and severity of climatic events affect both natural assets and human security; and accelerating changes such as of temperature and sea level rise affect the social cohesion of indigenous communities: in Alaska, for example, permafrost thawing and increased flooding are forcing villages to relocate.
>
> *(p. 194)*

Since the publication of this report, things seem only to have become worse. For example, while describing itself as "solutions orientated", in its scene-setting sections the subsequent *Global Environmental Outlook Report – GEO 6*, 2019 – notes that:

> GEO-6 comes at a time of great uncertainty about the current trajectory of global human development. . . . One major reason is that over the last few decades, human activities, such as human-caused climate change and other human impacts on ecosystems, have transformed the Earth's natural systems, exceeding their capacity and disrupting their self-regulatory mechanisms, with irreversible consequences for global humanity. . . . Humanity has already been seriously affected by ongoing systemic ecological changes, such as climate change and land use change (especially deforestation). These have reached the point that the ecological foundations of human society and natural systems that support other species and provide invaluable ecosystem services are in great danger.
>
> *(p. 4)*

More specifically, with regard to climate change, the report speaks of "time running out" to prevent the irreversible and dangerous impacts of climate change and its effects, including:

> extreme events (including flooding, hurricanes and cyclones) leading to loss of lives and livelihoods, pervasive droughts leading to loss of agricultural productivity and food insecurity, severe heat waves, changes in disease vectors resulting in increases in morbidity and mortality, slowdowns in economic growth, and increased potentials for violent conflict.
>
> (ibid. *p. 47)*

And on the matter of the effects of human behaviour on nature, it reports that:

> The increase in species extinction risks through time is well established, and there is no slowing in the rate of population declines globally. . . .

> Biodiversity loss is being experienced across all Earth's major biomes. . . . In the oceans, overexploitation of fish stocks is leading to fisheries collapse, warming is destroying coral reefs, and habitat destruction of coastal systems, such as mangrove forests, exposes communities to greater risks from erosion and extreme weather events. Marine plastic pollution is a major and growing threat to biodiversity. In freshwater systems, agricultural and chemical pollution, including increased nitrogen input, results in toxic algal blooms and a decline in drinking-water quality; invasive species are spreading through waterways; and freshwater species are declining at a faster rate than those in any other biome. In the terrestrial environment, rising temperatures are converting grasslands into deserts, and unsustainable irrigation has turned drylands into inhospitable, toxic landscapes unsuitable for wildlife or agriculture.
> (ibid. *p. 142)*

Amongst other things, this general situation has led to talk of a "sixth mass extinction event" and some prominent environmentalists are now in despair, anticipating an irretrievable situation in which equilibria that have enabled human flourishing over the past 12,000 years break down. For example, James Lovelock, proponent of the Gaia hypothesis that claims that there is an intricate multiple feedback system that maintains conditions favourable to life on the Earth, now speaks of the "vanishing face of Gaia" (Lovelock, 2009). While in some (now relatively few) quarters there remains debate over the extent to which some important elements of environmental degradation (such as climate change) are real, and if so whether they are anthropogenic, viewed overall, there can be little doubt that human action has a very significant – and ever-increasing – negative impact on the natural environment if we measure this in terms of the flourishing of the ecosystems that we inherited at the dawn of the Holocene – or indeed as recently as the dawn of the modern (industrial) age. Indeed, a statement published in *BioScience* on 5 November 2019 that was endorsed by some 11,000 scientists from 153 nations made it clear that in their view there is little doubt that human activity is having disastrous effects on the global climate. The declaration included the following statements:

> We declare clearly and unequivocally that planet Earth is facing a climate emergency.
>
> To secure a sustainable future, we must change how we live. [This] entails major transformations in the ways our global society functions and interacts with natural ecosystems.
>
> The climate crisis has arrived and is accelerating faster than most scientists expected. It is more severe than anticipated, threatening natural ecosystems and the fate of humanity.

Such is the impact of human activity on the global environment over the last two centuries that some commentators (following Paul Crutzen and Eugene Stoermer, 2000) now speak in terms of us entering a new geological epoch: the

"Anthropocene". This era, initiated by the Industrial Revolution, is one where humankind is taken to overwhelm important elements of the pre-existing natural order, hence exiting the largely benevolent (from a human point of view) Holocene period and entering a new era in which the development of the geological and biophysical environment increasingly is determined by human behaviour. It is held that this is an epoch in which the socio-ethical rather than simply the geo-biophysical determines environmental conditions: human decisions have become a geological force, leaving an irreversible imprint in the planet's rock and ice strata and atmosphere (Olvitt, 2017). For example, "technofossils" such as plastics and concrete, and unique concentrations of lead, nitrogen, phosphorous, and carbon dioxide associated with fertilizers and the burning of fossil fuels are now globally evident in anthropogenic deposits (Waters *et al.*, 2016).

The espousal of an Anthropocene era directs attention to three key features of our current situation. First, this is a period in which long-established natural processes, rhythms, and cycles are being progressively disrupted such that there is the potential for them to break down altogether. Hence, this is an age where old certainties can no longer hold, and in many respects new certainties – beyond that of radical change – are extremely hard to establish. Second, nonetheless, impetus is given to a perceived need for humankind explicitly to take responsibility for, and to attempt to manage, processes that formerly were determined through "natural causality". Third, so far human interventions in nature have frequently resulted in unforeseen and undesirable consequences, and, in overall terms, have led to the aforementioned widespread degradation. Hence, this reframing of the superordinate context for environmental issues in terms of an Anthropocene epoch foregrounds a number of interesting – and urgent – epistemological and ethical considerations. Broadly conceived, these include how to establish a serviceable knowledge base for understanding likely outcomes of decisions on a global scale and how to establish a global environmental ethic – both philosophically and practically. There are good reasons for believing that current (Western) ethical codes are not equipped to deal with this challenge (see, for example, Postma, 2006; Bonnett, 2012), and given the wide range of socio-economic, cultural, and geographical contexts to be accommodated, consensus over what is to be considered "right" and "good" in such an epoch will be difficult to achieve. As Lausanne Olvitt (2017) has noted:

> The sheer scale of human action and its intergenerational consequences in the Anthropocene marks a definitive shift in ethical life.

In sum, the idea of an Anthropocene era is useful in drawing attention to the scale, and to some extent the character, of the environmental challenges that now face us. Many of these will be considered in more detail in later chapters. But an important caveat needs to be voiced about allowing the idea of the Anthropocene to frame our thinking on environmental matters. While highlighting the global scale and significance, and likely irreversible effect of human activity on the environment, it is worth noting the towering anthropocentrism that runs through the idea of an Anthropocene epoch. It casts our perceptions of our current habitation in geological time in highly human-centred terms – maybe to the point of a self-aggrandizement that continues to set us above the importance of the processes of nature in which we are embedded. It will be argued that there are contexts in which this elevation of the significance of human agency can be deeply misleading and counterproductive.

At this point, it will be helpful to say something about idea of nature that is central to the themes to be developed in this book. The idea of the Anthropocene draws attention to both the idea of a geological timescale and the significance of human actions in nature. On a geological timescale, talk of "saving the planet" becomes otiose. Whatever we do or do not do, nature in this broad sense will continue. The climate will change, species will come and go (including very possibly *Homo sapiens*). Indeed, all life will eventually become extinct as, now on a cosmic scale, it seems that the Earth will become devoured by its dying sun. The point is that viewed on these grand timescales, everything is in a state of flux, and that is the natural order. While this perspective is sometimes useful as a curb on humanistic hubris, the sense of nature that informs the current work is chiefly that of the current state of the biosphere; furthermore, its underlying concern is with states of the biosphere that are consistent with human flourishing. To be sure, there are greatly differing interpretations of nature and human flourishing to be had within this broad perspective, and it is the task of this book to elucidate, refine, and evaluate some of the most important ones insofar as they have a bearing on education and its ecologization. But, given some of the criticisms of anthropocentrism that run through my account, it is necessary from the start to acknowledge this one fundamental sense in which all the views discussed are either implicitly or explicitly anthropocentric. Henceforth, when I speak of "anthropocentrism", I will be referring to it in some stronger or more exclusive sense than that of a concern for states of the biosphere that are in some broad sense consistent with human flourishing.

The failure of current policies: the case of sustainable development

Whether or not the idea of an Anthropocene era is helpful, it is now widely accepted that much environmental degradation has resulted from the burgeoning of modern societies that have adopted materialist models of development, and the three key points identified in the preceding section remain salient. In this context, it is understandable that some notion that purports to combine the idea of ongoing development with ideas of avoiding and redressing harmful environmental outcomes would be seen as an appropriate – and attractive – response to the environmental degradation that now besets us. Hence, the notion of sustainable development has arisen, and for over three decades it has played a central role in orientating responses to the growing acknowledgement of a serious environmental crisis. It will be helpful briefly to recap something of its history and character.

"Sustainable development" is a political term that was first introduced in *The World Conservation Strategy* (IUCN *et al.*, 1980) and perhaps was given its most influential articulation with the publication in 1987 of the report of the World Commission on Environment and Development, *Our Common Future* (Brundtland Commission, 1987). Here sustainable development was defined as "a development that meets the needs of the present generation without jeopardising the ability of future generations to meet their needs". This definition was widely taken up and was consolidated as an educational concern at the Earth Summit Conference held in Rio de Janeiro in 1992 attended by delegates from over 170 countries and whose centrepiece agreement was *Agenda 21* (UNCED, 1992). This included the proposal to introduce sustainable development (SD) into the educational programmes of signatory nations. Thus, it found its way (at least notionally) into the core curriculum of many nations, giving rise to the idea of education for sustainable development (ESD) that was lent further impetus by the decision of the UN General Assembly in December 2002 to launch the Decade of Education for Sustainable Development 2005–14. While further examination of these ideas will be the subject of the next section of this chapter, it is clear that both SD and ESD represent a policy for dealing with environmental degradation from the point of view of whatever are taken to be human needs, and in this sense they are essentially instrumental in character. And in seeking to bring about particular practical outcomes thought to serve human welfare, as they gain currency, they have the potential to affect educational practice on a broader scale than might initially be apparent. As this wider connection with education is an important part of the backcloth to understanding possible future directions for, and implications of, ESD, I would like here briefly to provide further illustration of some of the issues raised by this instrumental orientation.

First, in the context of education, a clear tension can arise between the focus on inculcating behaviours that are considered to contribute to sustainable development (such as recycling and energy conservation) and broader educational goals associated with developing the critical judgement and autonomy of pupils. For example, suppose that after due consideration, the personal judgement of a pupil leads them to reject the prescribed practices. What should be the school's response? It is understandable that the pressure of what feel like official ecological imperatives might push teachers/institutions into accepting behaviour modification as an overriding goal. But what will this response communicate about the school's conception of its underlying values and purposes, and the seriousness with which it respects the value of pupils' own reason?

Second, and related to the aforementioned, the instrumental motive present in Brundtland-type conceptions of ESD can incline us to emphasize a scientific-technological approach to environmental issues in which essentially nature is conceived as a resource to be manipulated and exploited in the pursuit of technological solutions. Here, the overall goal of SD appears to be clear, while the values that inform it remain implicit and unquestioned. A form of scientism arises that sits comfortably within – and also reinforces – an essentially authoritarian conception of education in which participants (both teachers and pupils) are required to adhere to "objective" goals determined by an instrumental rationality and to have their success accounted in these terms.

Third, and somewhat in conflict with the aforementioned, insofar as scientific (and other) approaches reveal the natural environment as a complex developing organic system, they suggest an integrated and emergent curriculum that is taken to reflect the organicism and holism of environmental issues. The precise kind of holism revealed will depend on the kind of thinking that holds sway and its portrayal of reality – for example, as ultimately an abstract/probabilistic system describable in mathematical equations, or a living flux of unique events in which purpose and agency find a place. But either way, rigid subject boundaries would seem to be inconsistent with communicating the nature of reality and hence appear as a serious obstacle to developing an understanding that could be the foundation for environmental action.

Returning now to the idea of sustainable development, after some three decades of being at the centre of policy, in many respects previously noted, our environmental situation has only deteriorated. How is it that this state of affairs could have come about? This question and contributing answers to it will be an enduring theme of this book, but for the purpose of sketching this general backcloth to the more detailed discussions that follow in later chapters, some preliminary points are worth making. Mainstream interpretations of sustainable development that follow along the lines of the Brundtland Commission definition have been subject to a range of criticisms.

To begin with, as an instrument of change, largely the idea of sustainable development has failed. Although still prominent in much of the rhetoric concerning environmental issues, at a cultural level it fails sufficiently to motivate us: despite all the scientific evidence – and granted some significant changes in behaviour – as a culture we continue in practices that seriously degrade nature and heat the Earth. Notwithstanding a recent upsurge of concern about some environmental issues, for many in the West, daily life lacks any feeling of oneness with the natural world in which our being is ultimately embedded. Indeed, in many respects, it seems fair to say that in general terms we are still behaving like irresponsible adolescents, so short-sighted are many of our actions. This empirical state of affairs is consistent with some philosophical criticisms that have been raised against taking sustainable development as an orientating idea for environmental education that will be sketched next. But first it is worth noting that in a cultural milieu perceived as being preoccupied with feeding and inducing an ever more extensive range of human desires, the idea of sustainable development can rapidly become equated with notions of restraint and sometimes feelings of guilt. In this context, it can be felt to emit a purely negative energy in which caution and protectiveness can become overweening such that nature morphs into a joyless realm of forbidden fruits. Richard Louv (2010) has documented how in the United States the building of tree houses, the collection of specimens, and the enjoyment of fishing (to give just a few examples) increasingly have become subject to legislative, health and safety, or moral strictures. In a bid to overcome a prevalent loss of direct contact with nature (what he terms "nature deficit disorder"), he speaks of a current need for the decriminalization of unstructured play in nature if we are to nurture new stewards of nature in future generations. A balance between protecting nature and encouraging an active engagement with it that develops a love and keen feel for its realities needs to be struck. But on what basis should this balance be struck? What kind of knowledge and what ethical guidelines would be appropriate?

Returning to philosophical analyses, perhaps the most damaging criticisms of "sustainable development" derive from that very combination of ideas that have made it such an attractive proposition to many: combining sustainability with the desire for development. For some, such a marriage is little more than oxymoronic, and if ultimately not this, a source of potentially severe tension because "sustainability" is taken to emphasize conservation in some sense and "development" in some sense the desire to have more, or better. In many cases, under the influence of global free market economics, this latter is seen in material/economic terms and involves setting nature up as a resource to exploit, which in practice frequently results in its depletion and degradation. One is tempted to think that a certain semantic sleight-of-hand is in play when the ultimately unsustainable goal of continuous economic growth (the Earth being in many ways a closed system for practical purposes) is apparently brought into harmony with a much vaunted eco-friendliness.

In addition, the ambiguity and vagueness arising over such things as precisely what is to be sustained and over what timescale, how needs are to be defined, and how priorities between them are to be adjudicated across a diversity of cultural settings allow such a wide scope for interpretation of the idea of sustainable development that it is possible for it to function as a banner for people of very different motivations and vested interests: for example, captains of industry seeking to sustain profit margins, politicians seeking to increase GDP and employment, and "eco-warriors" seeking to sustain natural habitats. At worst, in its questionable appearance of melding ecological limitations with economic development and social justice, sustainable development has been characterized as "an empty signifier which operates like a huge myth with pretensions of being a salvation narrative" (González-Gaudiano, 2006). Certainly, in the midst of this vagueness, there is a sense that mainstream politics is failing to face up to some crucial issues and is prepared to indulge mythical hopes.[1] For example, there is little or no acknowledgement that sustainability might necessarily incur negative economic growth and a lowered standard of living for many in the West. Quite the contrary, somehow the new "green" economy is portrayed as a "win-win" prospect in which economic growth and employment will increase with the development of new technologies and industries.

But perhaps this indeterminacy within the concept of sustainable development that allows varied interpretations and speculative hopes to blossom should not be viewed in wholly negative terms. Some have extolled the virtue of a relatively ill-defined conception of ESD precisely because such ambiguity can facilitate wider appeal. In its publication *Education for Sustainable Development*, UNESCO (2005) spoke of sustainable development as a "constantly evolving concept" that, nonetheless, includes as contributory elements "fostering peace", "fighting against global warming", "reducing North/South inequalities", "fighting against poverty", "promoting democracy" and "fighting against the marginalization of women and girls", and (finally) "being respectful of the Earth and life in all its diversity". While doubtless at the level of slogan, each represents a laudatory goal in its own right, when it comes down to practical implementation there remains considerable room for contestation and contradiction. Helen Kopnina (2012) has noted some of the debilitating consequences of a loss of focus on nature and the environment.

Others, for example, Robert Stevenson (2006), have argued that in the educational context, the meaning of ESD should be worked out by teachers as they address it in their teaching. He complains that

> [t]he policy discourse of ESD, like that of environmental education, has generally been constructed by policy-makers and academics as an abstraction decontexualized from contexts of practice and to be enacted by others, namely educators who have not participated in the formulation of the goals and concepts.

He argues that as the people who will teach it, educators do not need a concept or vision of ESD handed down to them, rather they need to construct their own understanding of sustainable development. A similar spirit informed the long-running OECD-funded Environment and School Initiatives (ENSI) project, with the addition that pupils were included in the constructing process, being encouraged to develop "action competence" by identifying for themselves, and seeking to address, local environmental issues in tandem with teachers and their local community (Elliott, 1999; Posch, 1999; Robottom, 2005).

The emphasis on teachers and learners working from and developing their own viewpoints is surely to be commended in education on the grounds of authenticity of understanding, but the considerable weight placed on the rationality of teachers and pupils as a source of discernment and wisdom needs to be considered. Stevenson allows that "non-rational factors" will be present and that they sometimes can override rationality, but implies that this can be addressed pedagogically through ongoing critical (rational) scrutiny. This, I think, raises a central theme that will colour the analyses to be developed in this book. It is this: essentially, the concern is not that rationality needs to be alert to and employed to expel or curtail non-rational factors – as if rationality were some pure impartial capacity or process – but rather that rationality *itself* has always been partial in the sense of pursuing particular goals such as to categorize and explain in the light of whatever purposes are current, and that in modern and late-modern times, increasingly these purposes have revolved around control and exploitation of the natural world. This can be understood as the operation in the West of a deep set of motives that install us into a particular version of reality. In other work (e.g., Bonnett, 2004) I have dubbed this "the metaphysics of mastery" (of which, more anon) and insofar as this is true, there is a critical need for ESD to alert thinkers to the motives that currently condition rationality itself – in effect, a metaphysical investigation – if the underlying meanings of ESD, current and potential, are to be revealed.

Meanwhile, in a time of the metaphysics of mastery, inexorably development acquires a highly Western economic interpretation that overrides the conservationist impulses present in the idea of sustainability. From this perspective, the "tail" of development comes to wag the "dog" of sustainability. ESD becomes enmeshed in the economism, scientism, consumerism, pre-specification, overweening managerialism, and standardization of language and procedures that are all instruments of a drive to control, exploit, and possess.

Good examples of how sustainability can become hijacked by this kind of mentality are provided by the increasingly proclaimed strategy of "carbon offsetting" and growing talk in academic circles of "ecosystem services" and "natural capital". On the account being developed, carbon offsetting in the form of, say, planting young trees that are taken to have the potential to absorb at least the equivalent amount of carbon dioxide that a particular human activity emits is misconceived in at least two respects. First, it embraces a dangerous "have now, pay later" principle in that the claimed beneficial effect will be delayed and in practice might not materialize due to lack of ongoing adequate cultivation. Second, and more fundamentally, it reinforces precisely key elements of a mindset that needs to be combatted. It offers to legitimate – and in this sense to normalize – the attitude of mastery of the natural environment, and also behaviour that is destructive of its integrity, by a practice that can be bought without too much personal inconvenience and that supports the illusion that the individual's responsibilities in this regard have been fully discharged. Similarly, the idea that the role and value of the natural environment can be adequately represented and safeguarded by commodifying it and measuring it by estimating in monetary terms what it contributes, or has the capacity to contribute, to the economy expresses in as consummate a way as is possible the attitude of regarding nature purely as a resource. Here the metaphysics of mastery reigns supreme and nature's myriad alternative countenances are completely effaced.

It is important to recognize here that none of this should be read as arguing that there are not times when we should attempt to compensate the natural world for the effects that our past and present actions might have produced, or that we should not regard nature as a supplier of our needs. Quite the contrary, as, in some important senses, part of nature, necessarily we use it and evaluate its potential in this regard. And in any case, in purely practical terms, the more the well-being of the natural world is understood to be closely allied to human well-being, the more we are encouraged to behave in ways that do not destroy it. What is in question here is not whether in some sense we regard nature as a resource, for this can hardly be avoided, but the underlying character of our interactions. For example, is all usage of nature of the same kind? Are there kinds that genuinely respect the integrity of nature – indeed, that can be understood as *expressing* this integrity? If so, how might this affect our understanding of economics and, perhaps, the place of aesthetic sentiments in deciding economic goals? These questions raise complex issues that will be explored in coming chapters. The answers that emerge will be central to the broader question of how we should respond to our current environmental crisis, and what is germane to the idea of ecologizing education.

Finally, I return for a moment directly to educational concerns of the kind broached previously. Notwithstanding the caveat concerning modern rationality noted there, the promise that the participative pedagogy commended by ENSI, UNESCO, and others has the potential to initiate fresh beginnings is also stymied by the influence of an underlying motive of mastery. Educational systems driven by pre-specification of outcomes and top-down managerialism hardly possess the spontaneity that is required to respond sympathetically to newly arising perceptions and opportunities as they concretely occur.

Post-ecologism and post-sustainability

That the exploitation and degradation of nature continue apace has been recognized and is powerfully articulated in some narratives that have arisen in response to the ascendance of sustainable development as a policy. A significant number of commentators have seen the latter as a rhetoric that cloaks practices that are highly unsustainable when viewed from the perspective of the natural world. This returns us to the question of the interpretation of sustainability that is being employed and the ways in which it permeates social practices. Two general avenues of response to this question have arisen: one descriptive and sceptical, the other advocatory and hopeful. The former has been described as "post-ecologism". This sees sustainable development discourse as an enduring pretence that reflects the economic interests of the West. The latter – "post-sustainability" – maintains the idea of environmental care as a powerful and real force, but translates it in a way that recognizes the flaws in the idea of sustainable development and that initiates a new aspirational narrative. This attempts to elucidate a deeper and more coherent idea of sustainability than that which currently dominates environmental discourse. Let me say a little more about each of these approaches.

Resonating with points made earlier in this chapter, post-ecologism refers to an analysis of Western culture that portrays it as essentially uncaring of the effects of commercial activity and consumerist lifestyle on the underlying character of the global economy, both natural and human. Behind a public simulation of deep concern for the state of the natural environment lies a real concern for economic growth and the rise of the material well-being of (largely Western) elites. This is facilitated by a rarely publicized transfer of deleterious consequences onto the globally poor, underprivileged, and weak. For example, toxic waste frequently ends up on their doorstep and they are often those worst affected by climate change such as extreme weather events, desertification, and rises in sea level. At the same time, globalized trading patterns make it hard for poorer nations to resist agricultural and economic practices that degrade their natural and cultural resources, leaving them ever more vulnerable to environmental problems. Expanding on his idea of the establishment of a "post-ecologist constellation", Ingolfur Bluhdorn (2002) contends that such phenomena as nature becoming cast as a heavily contested social construction and the "de-ideologization" of eco-politics and reformulation of ecological issues as economic and efficiency issues underwrite a situation where:

> we may conclude that the discourse and policies of ecological modernisation and sustainable development function to simulate the possibility and desirability of environmental justice and integrity without genuinely aiming to address, let alone reverse, the fundamental unsustainability of late-modern society.

Elsewhere, this has been characterized as a normalization of environmental problems, and is nicely illustrated by Bob Jickling's (2013) contribution to a symposium on this topic where he describes how in political negotiations contradictions in discourse between economic imperatives and the perils of climate change are smoothed over and reabsorbed into the status quo by reframing ecological issues in economic terms – for example, as issues of resource management and efficiency. He observes that: "In effect, this move absorbed differences, circumvented any serious discussion about values, and ensured that norms and assumptions of modernity and capitalism remained unquestioned, authoritative, and non-negotiable". For example, drawing on Blühdorn's (2010) commentary on the 2009 Copenhagen climate summit, he notes that adapting to scenarios of 4–7°C of warming by the end of the century – that formerly had been thought to be harbingers of catastrophe and disaster – became regarded as much more realistic than the tougher goals for decarbonization that the conference had been set to discuss. This shift is another illustration of a drift towards the politics of unsustainability – in effect the normalizing of catastrophe – that is often represented in the politics and rhetoric of adaptation.

In contrast to this rather bleak analysis, the post-sustainability movement attempts to retrieve and recast the idea of sustainability in a way that allows it a new vigour by demonstrating that there is a conception of sustainability that places it at both the heart of human well-being and the conservation of the natural world. As such, it holds out the possibility of both orientating a positive conception of environmental education and initiating an authentic conception of education as a whole. The arguments that support this will be explored in Chapter 4, but here it is important to note that in their different ways both post-ecology and post-sustainability views recognize and seek to combat the normalization of deeply unsustainable attitudes and practices. Philosophically, these attitudes derive from a long genealogy of ideas in Western culture that have set up nature as a resource to be manipulated and exploited. As these ideas remain current – and, indeed, in some cases continue to gain in strength – as a preliminary to further argument, it will be useful to give a brief account of them.

An unhelpful constellation: modernist humanism and the metaphysics of mastery

The development of the European Enlightenment brought to prominence a constellation of ideas that had a profound effect on our relationship with nature. I will argue that they lie at the beginning of a philosophical and cultural direction of travel that has strongly conditioned our current environmental situation. I rehearse three key elements of this Enlightenment thinking in the following (they are heavily interwoven and are here separated out only for the purpose of exegesis).

First, there was the elevation of a particular version of human reason. This construed understanding the world as being heavily linked to gaining power over it. It embodied aspirations and belief in the possibilities of subjugating nature to the service of human purposes. While motives of this general kind were not entirely absent in antiquity (for example, there is evidence to suggest that in the Hellenistic period, there was optimism over the ingenuity of man and his technology to shape nature [Glacken, 1967, pp. 117–120]), it was during the European Enlightenment that they became distilled and encoded in a particularly potent form that projected nature as essentially an inexhaustible resource. This portrayal of nature both conditioned the spirit in which it was studied – for example, Francis Bacon's recommendation, at the inception of modern science, that science wholeheartedly adopt utilitarian motives – and was conducive to the aspiration to make nature "on-hand" wherever possible. Martin Heidegger (1977) has described this latter as seeking to convert nature into a "standing reserve" that can be endlessly switched around in the service of human consumption. Here is revealed the final goal of anthropocentric mastery.

Second, and intimately bound up with the preceding point, there was the supposition that nature is to be adequately accounted in terms of a discursive rationality of humanly constructed categories and theories. In our exploration of nature, our stance should be that of interrogator who, according to Bacon, needs to "torture nature's secrets from her" (cited in Capra, 1982, pp. 40–41). Another luminary of the Enlightenment, Immanuel Kant, put it this way:

> Reason must approach nature not "in the character of a pupil who listens to everything that the teacher chooses to say, but of an appointed judge who compels the witness to answer questions which he has himself

> formulated. . . . [I]t must adopt as its guide . . . that which it has itself put into nature. It is thus that the study of nature has entered on the secure path of a science, after having for so many centuries been nothing but a process of merely random groping.
>
> *(Kant, 1933, p. 20)*

Clearly, this view heavily discounts the kind of tacit knowledge gained by an intimate living with nature – such as that celebrated by indigenous cultures – and the abstraction and idealization set in train is highly commensurable with supposing that, indeed, in essence nature is (merely) a human construction. In the hands of postmodern/poststructuralist accounts, nature perceived as a product of our categories, theories, narratives, and texts reaches its apogee. Maurice Merleau-Ponty (1962, p. 300) notes how, on this trajectory, reality becomes not something given in experience – "a crucial appearance underlying the rest" – but rather a superordinate "framework of relations with which all appearances tally". This is highly congenial to supposing that the most fundamental structures of nature can be articulated in mathematical terms, and this is precisely the project that was set in train by the thinking of Galileo. Today, it seems relatively unremarkable to suppose, as with the physical sciences, that the basic workings of the universe can be captured and explained by mathematical equations and that phenomena receive their most fundamental articulations by being assimilated to the structures that result.

Third, nature is understood as a realm set apart from the human – its fundamental constitution being conceived as purely physical matter/energy operating according to blind universal laws. Here we are presented with a mechanical causal or probabilistic system entirely innocent of internal purpose and meaning, and therefore incapable, in itself, of generating normativity or of possessing inherent intrinsic value. This positing of nature as purely physical is particularly overt in the thinking of Rene Descartes and expresses the deeper presumption that the structure of reality is to be supplied by the imposition of impersonal rational cognitive categories and ordering previously described rather than, say, through an affective sensibility that is open to aesthetic qualities, felt value, and worth. Frequently, such qualities emanate from a sense of the individuality and uniqueness of things in their sheer standing there. This is also eclipsed under the influence of abstract reason and instrumentalism in experience.

In the chapters that follow, I will attempt to amplify some of the ways in which this constellation of ideas plays into both a defective understanding of ourselves and a defective basis for decision-making on environmental matters. For the moment, I will simply make the general point that a fundamental motive running through this constellation of thinking is that of seeking mastery – particularly over nature. The Enlightenment goal of perfectibility is both humanly framed and to be achieved through human judgement and action. It seems to me that so culturally deep and conditioning of experience is this motive of mastery that it is legitimate to interpret it as a metaphysics. Its holding sway shapes our perceptions and ways of relating to the world so as to install us in – and installs in us – a certain kind of reality: one in which everything is turned into a resource in the service of the human will. Here the meaning and value of all that we encounter becomes a function of how it shows up in utilitarian calculations – for example, as having the potential to advance or frustrate the will, and in this or that particular manner.

The dominance of this metaphysics relates to another cultural phenomenon in the West: the growth of scientism. Again, a more detailed examination of this will be given in a later chapter, but it will be helpful at this point to give a preliminary characterization of what I mean by this.

Scientism in late-modern society

By scientism, I refer to the phenomenon of presuming that classical experimental science has a privileged access to the nature of reality – that somehow its methods, findings, and constructions reveal what is "really" real and that therefore it can assume the mantle of arbiter for thinking in general. Clearly, this is to be distinguished from science as a field of research; scientism is a set of presumptions about the significance and application of the assumptions, methodologies, and findings of this field of research in our daily lives. With regard to the natural world – which is here my central concern – it arises, for example, in claims that what in everyday experience we take to be solid objects are to be understood as, say, "really" bits of space traversed by speeding particles; what we experience as their colour or sound is "really" movement of a particular wavelength. When it appears to us that a beaver selects a site to build its lodge, protects this site from river surges by quiet pools resulting from felling nearby trees, gnawing them to manageable size and towing them to narrow parts of the river to construct dams, what is "really" occurring is the working out of blind mechanical processes. The vocabulary of the former everyday account in which purpose and agency are attributed to aspects of nature is to be regarded fundamentally as a quaint piece of anthropomorphism.

When operative in the social world, scientism can be equally stringent. Take the example of education. Here scientism expresses itself in the requirement for teaching to be organized around a detailed pre-specified curriculum with outcomes that can be objectively measured through public tests whose results can be expressed mathematically, and that can be compared across a range of situations. In terms of curriculum content, it recommends the development and application of skill sets that, again, can be detached from particular learning contexts and can be widely applied. Hence, for example, there is strong encouragement to identify and teach various "higher order" cognitive skills such as "analysis" and "problem-solving" that it is held can be deployed across a range of contents, turning things encountered into so much data that can be manipulated and mastered by the human intellect.

What is wrong with this orientation? Perhaps nothing if it remained contained, did not seek general dominance, presents itself as the chief – even exclusive – goal of education. But today frequently it aspires to precisely these ambitions, and in the context of education there is a high price to pay for their achievement (especially with regard to achieving a full engagement with nature): the sanctity of the particular – the individual in its singularity – is annihilated. Its being is reduced to that of instance of a general category and the richness and immediacy of experience in which unique vibrant things communicate with us is lost. On this account, rather than providing a privileged access to the reality of nature, scientific narratives have the effect of effacing important aspects and, indeed, if allowed to dominate our thinking, of subverting its authentic significance for humankind. Perhaps this is acceptable within the discipline of science itself when its limitations are acknowledged and accepted as a price worth paying for achieving a particular kind of objectivity. But when scientific presumptions, methodologies, and pronouncements become generalized beyond the discipline and assume the countenance of arbiter of thinking and understanding more broadly, great damage is done. Unfortunately, this generalizing of scientific presumptions is pervasive. It sets the tone for many activities that lie beyond its proper scope, including, as I have noted, that of education. And as part of a broader power play that I have dubbed a metaphysics of mastery, it is complicit in a destructive alienation from nature.

I will further substantiate and explore more fully the implications of this claim in a later chapter. For the moment, it is worth noting that many commentators today argue that our relationship with nature is in a parlous state. They report on what they see to be a worrying disconnect with nature. There is discussion of the growth in Western society of a kind of autism with regard to nature, of "plant blindness", of a "bubble wrap" generation insulated from nature and caught up in hyper-consumerism and neophilia that results in large-scale effacement of the reality of the natural world. Richard Louv (2010) has articulated much of this and its extensive consequences in his exploration of what he has termed "nature deficit disorder", and David Abram has elucidated some of its phenomenological antecedents: for example, the way that language, with the advent and spread of phonetic literacy, has become divorced from the "flesh" of speech and the "flesh" of an animate world (Abram, 1997, chapters 3 and 4). Taken together with the post-ecology arguments previously mentioned, it can be seen that there is a growing body of literature that detects the normalization of both deeply alienating attitudes towards the natural world and a manufactured oblivion to the environmental catastrophe in which we are increasingly heedlessly embedded. I will develop the view that insofar as this is true, it is destructive not just of our material prospects but also of our spiritual well-being.

This brings me to what is perhaps a central, but frequently overlooked, question concerning the topic of environmental catastrophe: *what is the real catastrophe?* In line with the claims made previously, I will argue that it is that in dominant strands of Western culture we do not, and seemingly cannot, think properly about the issues involved. While environmental degradations of increasing intensity rightly now command our attention, too often there is little acknowledgement that these are the symptoms of a deeper malaise: a discursive scientism that, under the aegis of the metaphysics of mastery, too frequently has become the default mode of thinking when it comes to addressing environmental (and many other) issues.

This is not to say that, for example, within the field of environmental education research, there have not been powerful critiques of presumptions of a prominent place being given to techno-scientific approaches and artefacts. Taking the latter, Phillip Payne (2006) has developed a phenomenological perspective that he takes to reveal highly problematic effects of incorporating computer-mediated experience into environmental education because of the impoverished quality of perception and bodily engagement that it frequently entails. This is an important consideration, to which I will return. Others, such as Ian Robottom (2005), have discussed the way in which during the 1970s and 1980s, an applied science paradigm dominated research in environmental education – and the subsequent attempt to establish an approach deriving from the emancipatory tradition of critical theory, as exemplified in the previously mentioned ENSI project. Here, instead of environmental education being conceived as the implementation of a centrally devised objective scientifically factual curriculum, students, teachers, and communities become true participants in curriculum development by taking responsibility for identifying local environmental issues and exercising their own critical rationality in analysing them and addressing them in practical terms. Here a traditional separation of education and research is transgressed: the participants are the researchers, generating properly contextualized knowledge rather than simply being receivers of abstract objective knowledge produced by others, and that pretends to universal application (see, for example, Elliott, 1999; Posch, 1999).

While clearly in some respects these examples demonstrate that there are significant pockets of resistance to an overweening techno-scientific paradigm, two things need to be noted in the context of the argument that I wish to mount. First, the impact of such approaches on the educational policies of Western national governments remains limited. Second, with respect of the participatory research approach, as previously pointed out, a strong possibility remains that the rationality used by participants to determine and analyse the environmental issues that they identify will reflect the heavily instrumental and anthropocentric stance that I have taken to pervade Western society at large. The extreme partiality of perception and understanding of nature that this deeper cultural framing involves makes any thinking that is held in its sway inadequate to the task of making properly informed environmental decisions. The character of our quotidian engagement with the world/our environment – and how a particular reductive kind of thinking has become normalized – therefore becomes a critical question.

Some key themes

The foregoing survey and preliminary analysis of some central issues posed by consideration of our current environmental situation provokes a number of questions and themes that it will be the business of subsequent chapters to pursue. I list them as follows:

1 What would constitute a proper account of nature that restores to it those qualities that scientism and the metaphysics of mastery have occluded? Here, I will develop a phenomenology of nature in which, amongst other things, its transcendence and inherent agency, normativity, and intrinsic value are displayed.
2 What, philosophically, are the key obstacles to a full engagement with this revivified experience of nature? Here, I will elaborate on some aspects of modernist humanism and scientism that I take to be particularly salient in the effacement of nature and that lead to our estrangement from it. This will include a discussion of the erosion of the significance and authority of nature in intellectual life and the dominance of mathematicized linear time in quotidian life.
3 Is there an essential relationship between human being and nature? If so, how is it to be explicated and what are the implications for authentic education? Elaboration of this theme will include development of the idea that education itself, properly understood, is intimately concerned with an individual's being in the world, and therefore is ineluctably environmental. This is supported by an account of the intentional – and therefore ecstatic – nature of consciousness. Furthermore, it is argued that a central dimension of this environment with which ecstatic human consciousness is engaged is that of *nature* understood as transcendent in the sense previously outlined. It will be held that such engagement with nature presents opportunities for consciousness quintessentially to go beyond itself, to be inspired and refreshed, and to receive non-anthropogenic standards in the form of intimations of what is fitting and what is not.
4 Given this conception of human being as essentially environmental and therefore in principle open to nature's address, are there implications for mainstream social and moral principles? I will argue that because of their inherent anthropocentrism, current mainstream conceptions of social and moral principles are revealed as cabined and inadequate to the task of addressing our current environmental predicament. I will take as an exemplar of this claim an examination of contrasting ideas of social justice and "ecological justice" in the context of what I argue to be a key and perennial issue: human population growth.
5 How does the character of ideas central to education such as truth and knowledge appear when viewed from the perspective of environmental consciousness and the phenomenology of nature previously developed? What credence can be given to ideas of nature having a "voice", and how, if at all, would this contribute to the development of the kind of systemic wisdom that would be central to an environmentally responsive education? In this regard, what issues are raised by the burgeoning of digitalized experience?

6 Is there a case for the ecologization of education and the philosophy of education? If so, in what sense and what would be involved? Here ideas such as developing a sense of wonder and the value of immersive experience in nature will be reappraised. Also, I will explore how recognizing nature's alterity, integrity, and intrinsic value revealed in its emplaced presence transforms mainstream ideas of moral and social education, and how nature's fragility and possible demise (when referenced to the current state of the biosphere) require a discussion of education for a future that is not certain.

In the introduction to his influential book *The Nature Principle* (2012), Richard Louv wrote:

> because of the environmental challenges that we face today, we may be – we had better be – entering the most creative period in human history, a time defined by a goal that builds on and extends a century of environmentalism, which includes but goes beyond sustainability to the re-naturing of everyday life.
>
> *(p. 5)*

In elucidating this, he goes on to speak of idea of the Nature Principle as "the power of living in nature – not *with* it, but *in* it", and of the twenty-first century as the "century of human restoration in the natural world". Amongst other things, this raises the need to consider "re-naturing" the places where we live such that:

> The question of human/nature kinship is one of the great architectural, urban planning, and social challenges of the twenty-first century.
>
> (ibid. *p. 135)*

In order to evaluate both the possibility and the desirability of this kind of aspiration (for given the influence of the metaphysics of mastery, it cannot be assumed that "getting back to nature" will appeal to everyone – at least not immediately, nor can it be assumed that there will not be serious – indeed, trenchant – practical, political, and economic obstacles to achieving it), it is necessary to engage fully with the questions listed previously. But in addition to these questions, a further fundamental one is now revealed: what would it mean to live *in* nature? Presumably something more than simply being located physically in a natural environment is implied, for one could be so located and yet have little interest in it or respect for it. Perhaps an active acknowledgement of kinship with nature and participation in its processes is required – some sort of immersion in which feelings of belonging and "oneness" arise. On the other hand, to become fully immersed in nature – as it were, simply to become a part of it and no more – is also problematic. In many ways, nature is, as the Victorians had it, profligate, "red in tooth and claw", perhaps governed by laws such as the survival of the fittest. To be carried along in this way would be a denial of many fundamental human values. If we must participate in nature, it must be *as* human beings.

Hence, it is the character of one's "being in" nature that is important – the quality of one's relationship with its salient features. And this, in turn, will involve one's conception of nature – what one experiences as, or takes to be, its essential qualities and the kinds of response that these require. These are deep philosophical/ethical issues that require careful elucidation if we are to grasp properly our current environmental situation and what it demands from us. It is becoming overwhelmingly clear that addressing issues such as climate change is a matter of urgency and developing practical means of reducing carbon emissions clearly is highly desirable. But the full implications of these and similar issues can only be properly understood if the extent and character of the normalization of catastrophe is addressed.

Note

1 At the time of writing, the frustration with the lack of political action with regard to the current environmental crisis has erupted in the public demonstrations of the "Extinction Rebellion" movement. This movement has three key demands:

   1 The government to declare a climate and ecological emergency and to work with other institutions to spread a message calling for change;
   2 The UK to "act now" to stop loss of habitat and to reduce carbon emissions to net zero by 2025;
   3 The government to set up a Citizens' Assembly made up of people across society to decide how to solve the climate crisis with advice from experts.

   (Reported in the *BBC Newsround*, 15 October 2019)

2

A PHENOMENOLOGY OF NATURE

The "occurring" of things in nature

In Chapter 1, I briefly alluded to some cultural influences that incline us towards a cabined and distorted understanding of nature and hence a deficient relationship with it. This is a matter of great concern because, on the view being developed in this book, it is our underlying attitude towards – and understandings of – nature that lie at the heart of our current environmental problems. The ambition of this chapter is to begin to unfold a more revealing understanding of nature that restores to it essential features that were dismissed or overridden by Enlightenment attitudes and the burgeoning of a metaphysics of mastery. By way of entry into this complex terrain, I begin by considering two ways of viewing the now widely accepted idea that things in nature exist always in reciprocal relationship and how these different viewpoints affect our experience of nature and our understanding of our place within it. The first of these orientations is that of scientific ecology, which, based on the rationality of the natural sciences, became very influential from the middle of the last century onwards and has formed a cornerstone of much subsequent "green" thinking. The second perspective is phenomenological. This also has been incorporated into some green thinking, but on the account that I give invites a somewhat different underlying orientation when compared with scientific ecology.

Two views of nature

Taking the stance of scientific ecology, the interrelationships detected in nature ultimately are construed in terms of causal or probabilistic law-governed biophysical interdependencies. Individuals are perceived as causally sustained as integral members of local ecosystems, which in turn are nested in overarching regional or global systems. Hence, from this perspective, an area of woodland might be described in terms of its geographical (including climatic) location and as a habitat containing a number of definable entities, such as mathematically accountable populations of particular species of flora and fauna whose biophysical interdependence can be mapped objectively in various ways by an impartial observer. For example, the tree roots stabilize the soil, their foliage provides shade, and eventually contributes nutrients and structure to the soil as it is incorporated by the activities of particular species of insects, earthworms, fungi, microbes, etc., which in turn depend on the leaf-fall for their own survival, and so forth. Undoubtedly, this is all correct. Accounts of this kind make the invaluable point that nature does not consist of isolated entities, and that anything that impacts the biophysical functioning of any individual will impact the rest of the system of which it is a constituent part – sometimes in very complex ways and over very extended periods of time.

It is hard to overstate the significance of this simple point when it comes to evaluating the effects of human activity on the natural environment. And it has to be acknowledged that because of the geographical and temporal extent of natural systems, the full nature and extent of these effects can be hard to anticipate. The history of even well-intentioned interventions in nature is littered with unintended consequences. A classic account of an example of this is given in Rachel Carson's *Silent Spring* (1962) where the disastrous effects on wildlife of the use of DDT for the purposes of pest control on food crops are documented. Eventually, the agricultural use of DDT was banned globally in 2001, but the dire consequences of it entering complex natural food webs are still with us, particularly in aquatic situations where its half-life has been estimated to be 150 years and, for example, due to the process of global distillation, it has become concentrated in Arctic regions with deleterious effects on apex predators. It is for this reason that protagonists of sustainability attach great importance to the "precautionary principle" that requires that the onus is put on those proposing projects that affect the environment to demonstrate that they will not have harmful consequences somewhere within the greater system.

Clearly, this kind of system-orientated thinking is of great value, but it needs to be noted that it emanates from an essentially partial engagement with nature and that the understanding that it provides is therefore incomplete. This is because it arises from an essentially disembodied encounter with nature in the sense that it does not linger with individual things in their sensuous and aesthetic being. Rather, it conceives things exclusively in terms of the functions that they serve as members of the systems in which they are taken to operate. This preoccupation with causal connections between categorized objects located in abstract systems leads it to discount the significance of visual, acoustic, olfactory, and textured phenomena as they are immediately experienced. Because of this, fundamental aspects of natural places are rendered invisible by this perspective. While, undoubtedly, providing important insights into the natural world, this effacement of it in its fullness that scientific ecology involves becomes highly significant when it comes to understanding how we should relate to it.

With this in mind, and by way of comparison, *phenomenologically*, things look somewhat different. Instead of a place being conceived as a particular type of habitat functioning through the operation of blind physical laws observable by an impartial spectator, natural places become unique sites of intelligibility that address us as individuals, each presenting its own countenances and ambiences. And the very occurring of each individual thing within a place both conditions, and is conditioned by, it. Harking back to the previous example and refining the focus on what for present purposes I will term a "more natural" place, consider the experience of entering a woodland dell in late spring. In such an encounter, we can become aware of another important – and I argue more fundamental – sense in which the things of the woodland dell are interdependent. They are dependent on each other not simply for their physical survival within a causal nexus, but in their *being*. By this I mean that their very occurring, their way of standing forth as the things that they are – is a matter of mutual sustaining. Let me elaborate on this a little.

When we speak of becoming aware of someone's presence, we can refer to an awareness of them that somehow transcends or undercuts what cerebrally we might know of them, what constitutes them categorially, as it were. We can *feel* their presence. Similarly, on our arrival at the woodland dell, should we encounter, say, a great beech tree we can feel ourselves to be in its living presence. We can experience it as standing there in and of itself. This remains true even if we do not know exactly *what* it is, that it is a *beech* tree. For example, we can be struck by the inscrutable massiveness of its gleaming boughs, yet that dance in the breeze; the fall of its extending and diminishing shadow on the vegetation below; the rustle of birds flitting through its foliage, itself delicately intermingled with that of its neighbours at the periphery; the aggregations of moss and lichen on its limbs; the enveloping odours of growth and decay; the dance of midges beneath its canopy as evening closes; and so forth.

Here it can be seen that the manner in which this particular tree occurs is imbued with the occurring of all that is around it, all that shares this particular place. By participating in this interplay, the tree both contributes to sustaining the place in which we come across it and is sustained by it. Through its unique and infinitely manifold contribution to the precise ambience of its neighbourhood, it upholds this neighbourhood – contributes to the unique and ever-changing qualities of its space – and is upheld by it. In so doing, it participates in a *place-making*. In concert with its neighbours, it radiates a sense of the particularity of *this* place – what it is to be *here*, as distinct from anywhere else. To borrow from Gerard Manley Hopkins' poem *Binsey Poplars* in which he speaks of "the sweet especial scene" whose beauty after-comers cannot guess once the trees are felled, what is "especial" about this place fills the air, and this also conveys a sense of what *belongs* here and what does not. Hence, physically extracted from this especial place, there is an important sense in which the beech tree is no longer the same tree. It cannot occur as once it did; part of its original "selfhood" has been destroyed (Bonnett, 2009b, 2012). Somehow transplanted to, say, a city mall, its evening being is transformed from a sheltering presence, to, perhaps, a cold silhouette on neon.

Of course, this is not to deny that in the perfectly straightforward sense of its material or biophysical continuity, the transplanted tree remains the same tree. It is simply to make the claim that in another, perhaps more important, sense it does not. And this illustrates a key point concerning the phenomenological approach. While scientific knowledge of the natural world can (rightly on occasion) colour how we experience it (for example, knowing that a species of mosquito transmits malaria to humans is likely to condition our attitude towards it), phenomenology is receptive to facets of experience that go beyond what we take to be the purely physical facts. For example, it embraces intuitions of ambience and mood, well-being and distress, and sensations of beauty and discord. What we might now term "phenomenological ecology" encounters the world in the fullness of its many experiential facets and seeks to discern and articulate ontological themes that emerge through such experience.

In relation to our understanding of the natural world, this illustrates the importance of the issue of the kind of thinking that should inform it, and in this case foregrounds two metaphysical approaches that in other work I have dubbed the "metaphysics of objects" and "metaphysics of things" (Bonnett, 2004, p. 65). The former sets everything up as definable in terms of the characteristics of category membership – as when, say, a tree is identified as being of a particular species according to a set of defining objective properties such as the shape, colour, and distribution of its branches, leaves, flowers, and fruit, its mature height under normal conditions, etc. Thus, fixed as a defined object, it can be stored in a database; its being is stabilized, organized – made ready for inspection, calculation, and use. In this way, it is intellectually processed and possessed. Essentially, it is on this conception of things as definable objects that scientific ecology operates. The holism that it perceives consists in systems of intimately causally connected, defined, phenomena (organisms, geological components, forces, interactions, and so forth) that in principle can be fully specified and apprehended by discursive reasoning. Their existence and interrelationships are a product of logical calculation. By contrast, the metaphysics of things does not see the tree as knowable and capable of summation in this way. It is alert to the tree's individual living presence, along the lines previously described. As the senses become attuned to the tree's address, the myriad aspects of the "thisness" of the tree become apparent and defy totalization under some classificatory label, as does its sheer existence. The relationships of these aspects with each other are not logical, but organic in the sense of consisting in mutually sustaining fluid interplays, and they are experienced as spontaneous, ineluctably "other", and hence inherently mysterious. They are known (always incompletely) by direct acquaintance rather than by calculation.

To take one more example: an upland stream. Here our attention might be caught by the glistening flow of water eddying around tumbled polished stones; the mysterious movement of a reed at the margin where the water is quiet; the momentary silver glint of a darting fish; and the fresh breeze that blows at this place and that disturbs the pendant branches of a stunted willow that overhangs the stream, its fissured bark displaying and withholding strange shapes as sunlight and cloud shadow pass over it. Here, again, the living presence of each is sustained through its participation in a creative interplay with all, and a unique place is constituted. Removed from this place – perhaps the stone to a rock collection, the fish to an aquarium – their being is transformed, reduced to that of curio or an item of decor. In such uprooting, their existence arises through their participation in a new imposed interplay that is in part both parasitic on what they once were and at the same time subverts key aspects of their original nature, their ability to befall us as natural.

What this phenomenological ecology reveals is that central to our experience of things in nature are qualities of deep organic interrelatedness and emplacement. These qualities are to be understood not as merely contingent but also as *ontological* in the sense that they constitute things in their *being*: their ability to stand forth in our experience as the things that they are. Hence, they are more fundamental than the relationships that form the focus of scientific ecology: for things in nature to be specifiable and locatable in, say, causal networks, they have first to stand forth. This holds true even when they first come to our notice in the course of exploring such networks.

Implicit in the phenomenological account of nature being developed here, there are some further important qualities. In the next section, I outline their character as a prelude to further development and refinement in subsequent chapters. But first it will be useful to consider a reservation that might arise concerning the treatment of nature that is now underway.

The reservation can be expressed as follows. How plausible and appropriate is it to continue to speak of nature in terms of it being one general sort of thing? Does this not carry the danger of homogenizing something that is manifestly infinitely diverse in its occurring? Surely these differing manifestations require differing kinds of understanding and responses? For example – and at a very basic level – whether we are confronted by a zebra or lion, a magnificent sunset or an outbreak of Ebola might (rightly) evoke very different responses. It might be said that to lump such different phenomena together as one "nature" frequently is simply misleading. Also, from a different standpoint, this aggregation of nature might be seen to invite the reinstatement of that disrespectful Enlightenment attitude towards nature noted in the previous chapter. Here the individuality of members of the natural world is reduced to some amorphous subhuman, purely physical, "other" that can possess no intrinsic integrity or worth. These are very valid concerns. Clearly, there are many instances when careful attention to the individuality of natural things and phenomena is necessary. Indeed, the account currently being developed lays great stress on this very point. However, to grant this is not to invalidate the current attempt to identify key aspects of experiencing something *as* natural. Indeed, the individuality of particular things in nature can hardly be voiced without recognition of these key aspects, and therefore is only enhanced by the account that is in progress. Hence, while remaining alert to the dangers that have been outlined, I return now to an account of nature, *qua* nature, that seeks to reveal its transcendent qualities.

Transcendent nature

While not wishing to deny our ability to discriminate within the natural world – and the frequent need to do so – I would like to suggest that for present purposes, there is some virtue in maintaining a holistic sense of nature – but *not* now only in the prevalent ecological sense of a vast interconnecting system in which human beings are nested, but rather in the sense of some key *qualities* implicit in our experience of all things that we perceive as natural. I believe that these are disclosed in the vignettes given previously.

One key quality that these vignettes display is that in our experience of what they portray, we come across natural things as standing there independent of us – i.e., as pre-eminently having their *own* being that we can affect, but of which we are not the author. This is to say that we experience things in nature as "*self-arising*".[1] It seems to me that this essentially non-artefactual quality of the standing forth of a thing from out of itself is a definitive feature of our experience of nature, whether it be a star, a mountain, or an amoeba. This is not to say that there are not many occasions when we will have played some part in the appearances of nature – for example, we plant trees, maintain or destroy habitats, invent instruments that allow us to perceive nature in new ways. These all influence the occasions on which, and situations in which, we encounter things in nature. But in their natural aspect, their nature *qua* nature, things in nature radiate a particular autonomy. They lie ever before and beyond our intentions, and while we can *affect* them in all manner of ways, we do not ultimately *determine* them in the sense of assuming sole authorship of the occurring of things.

Consider, for example, the case of our own bodies. Clearly, these can be affected by our choices and actions, but we maintain (or destroy) our health by interacting with processes of which ultimately we are not the progenitor and that, while we can sometimes influence their progress, have their origins in a realm that lies beyond anything that we plan or construct. This is to say that there is a natural order that we recognize as external to our will and with which we have to find a harmony, or at least reach an accommodation. The key point here is that in this sense nature *qua* nature is transcendent, "other". And in more general terms, we construe things in nature as leading their own lives – as with a flying insect disappearing into the warm dusk air. Indeed, we experience them as not standing forth exclusively for us, but for other sentient beings – in myriad ways and on myriad occasions, many of which we will never know of, and of which we can barely conceive. As an earthworm burrows through the soil, who can guess the particular ways, and occasions upon which, its presence has affected other inhabitants?

Considerations of this kind relate very closely to a second feature of nature *qua* nature: it is epistemologically mysterious. Although there may be occasions when we feel ourselves to be closely involved with an aspect of nature – and as a result feel that we know it intimately – as self-arising, nature is that which can never be *fully* known, intellectually possessed. As demonstrated in the earlier vignettes, nature is composed of (expressed in) open, multifaceted *things* whose occurring is fluid and ever-changing. Each has its own unique history and draws towards its own open future. It is capable of exhibiting an infinity of profiles and countenances, only some of which we will ever witness. This contrasts with previously mentioned defined *objects* of thought, whose being is exhausted by the characteristics of category membership that we devise and impose. (Recall the reduction of the living presence of the beech tree that occurs when it is summed up in terms of static objective properties that can be entered into a database, previously discussed.) Such typological descriptions, along with generalized causal explanations and scientific "laws", say nothing about the *sheer existence* of natural things. They give no insight into the experience of their individual standing forth in their suchness and their ability to affect us in unique and never wholly predictable ways. As one commentator put it: "Surprise is the general reaction of the attentive walker in natural space" (Grange, 1997, p. 96). And the appropriate epistemological response to self-arising nature then is not to seek summative knowledge, but to allow a sense of the ineffable.

It seems to me that the thinking of the Italian post-Enlightenment philosopher Giambattista Vico is pertinent here. He argued that we can only have full knowledge of things of which we are the author, for those who create something can understand it, as mere observers of it cannot. Thus, insofar as we actively participate in our own development, we are capable of having a more definitive understanding of ourselves and the human world than of nature, because we have an insiders' view. We know what it is like to *be* a human being in a way in which we cannot know what it is like to be a lion or a tree. With these latter, there remains a sense in which to a significant degree we are merely passive observers looking on from outside, capable only of "dark speculation" about the inner lives or goals of what we see (and if indeed, there are any such goals); capable only of seeing the surfaces of things and events that are essentially alien and mysterious. As Berlin (2000b, p. 13) puts it, for Vico the laws of the natural world are *knowable* but not *intelligible*. Thus, for example, one can know of the beliefs of another person or the practices of another culture, yet they can remain largely unintelligible until some further explanation is given, communicated. And even with this one may fail to achieve the kind of "internal" understanding to which Vico refers. And, of course, in the case of non-human nature, no such direct communication has yet been forthcoming. On this kind of account, the anthropomorphism that is so ubiquitous in our everyday perceptions of nature (particularly those that we offer to children) is seen to destroy its sheer otherness.

This point is both serious in its own right and because of a consequence that flows from it: a proper moral attitude towards nature cannot simply be some kind of extension of a human ethic as is sometimes proposed (see, for example, Peter

Singer's (1993) attempt to extend utilitarian ethics to the natural world). Not only are things in nature not the rational agents that we take ourselves to be and that is a foundation of many mainstream ethical positions, in many key respects their "good" is nothing like our "good". In many (but not all) regards, the natural world and the kind of normativity that pervades it is far removed from the human world. For example, the idea of an individual's responsibility for their actions, if applicable at all in the natural world, has a very different meaning and ethical significance. The actions of human beings are capable of being influenced by reasons rather than simply causation. This is to say neither that our actions are always guided by good reasons, for we are capable of irrationality, nor that there are not underlying causes that might be uncovered. But, as Richard Peters (1959) once pointed out, in the human sphere, we should not equate causation with unavoidability. For example, in the case of an action for which no plausible reason can be given, we might follow Freud and surmise that it was caused by some repressed event in childhood. But now knowing this, we can be put in a position to do something about our behaviour. Our choice and personal responsibility is restored to us. And, indeed, as Peters notes, we normally only seek out some cause to explain an action when a plausible reason for it cannot be found. But, as far as we know, there is no strict equivalent to this in nature; there are only causes. If we were to speak of a lion being responsible for the death of a Zebra, it would only be in the sense that the lion was the biophysical cause of the zebra's demise, not that it was in anyway morally culpable.

A parallel argument applies with regard to the place of pain. In the human world, the Benthamite equation of pleasure with good and pain with evil, and making this the basis of an ethical position, has some plausibility. But in the natural world, any such plausibility evaporates entirely. As J. Baird Callicot (1995) has observed:

> Pain and pleasure have nothing at all to do with good and evil if our appraisal is taken from the vantage point of ecological biology. . . . An arctic wolf in pursuit of a caribou may experience pain in her feet or chest because of the rigours of the chase. There is nothing bad or wrong in that.

He points out that in animals, pain informs them of bodily stress, irritation, or trauma, which is essential for their survival. For example, a certain level of pain can be a desirable indicator of degree of exertion during a chase, or injury to which an animal needs to respond: "A living mammal which experienced no pain would be one that had a lethal dysfunction of the nervous system". Furthermore, in the context of the natural world, an ethical aspiration to minimize pain would seem to betray a "world-denying or rather life-loathing philosophy". In many respects, the natural world, as actually constituted, is one where one being lives at the expense of others and pain is simply an integral part of this system. To seek its elimination is biologically preposterous, and if achieved would be ruinous. And of course it is entirely otiose to condemn the killing of a prey by its predator.

On the account being developed, such cautions against supposing that we can know nature from the "inside", or that we can properly project human values upon it, are well made. In more general terms, they alert us to a form of arrogance that has often led to misplaced actions – and sometimes disastrous consequences, such as when we have attempted the biological control of pests by introducing alien predatory species that become "invasive". No doubt, genetic engineering is susceptible to this concern and perhaps threatens dangers of a greater magnitude. Attempts at large-scale climate manipulation are the stuff of disaster movies. Yet, while in the era of a metaphysics of mastery hubris of this kind is always a danger, when we genuinely open ourselves to nature as the self-arising, we can – and often do – have a sense of what would count as the well-being, even interests, of things in nature. Indeed, Law Professor Christopher Stone (1974) argued that it is at least as plausible to allow that "natural objects" have interests that can be legally represented as it is make this claim on behalf of a comatose person or a corporation. While necessarily tentative, this possibility has important implications that will be explored in a later chapter when the idea of nature's having a "voice" that issues an ethical challenge to us is explored.

For the moment, it can be said that although they are profoundly other, things in nature can communicate something of their own integrity such that we can have some sense of what would count as their fulfilment. In this sense, they are normative and possess intrinsic value. Negatively, our awareness of this can be evoked if, perhaps, on entering the woodland dell previously described we were to find it strewn with the remnants of fly-tipping, or the beech tree to be wantonly vandalized. Or, returning to the upland stream, we were to find the bloated corpses of fish borne on foam-topped waters smelling of industrial waste. More positively, in experiencing the myriad interplays, harmonies and contrasts, and subtle adaptions and accommodations that we find there, we might be struck by a sense of *rightness* emanating from the dell or the stream during our first encounter. Somehow things occur in such a way that how they *are* communicates that this is how they *ought* to be. To make this point is not to argue that there will not be times when it seems wholly justifiable to disrupt nature's integrity in the service of human welfare. Rather, it is to argue that such integrity is present and is relevant to any decision to do so.

Furthermore, in part, this sense of rightness that we can experience in a natural place arises from the revelation of another central aspect of nature: things in nature participate in the play of transcendent elemental powers that run through and enliven the phenomenal cosmos (i.e., the cosmos in its appearing and standing forth). I have in mind here such powers as birth and death, growth and decay, lightening and darkening, sound and silence, motion and stillness, revealing and withholding. Their rhythms and interplays light up consciousness; they make a silent call upon it that carries it into an infinitely extending and mysterious universe where the known constantly rubs shoulders with the ineffable. Our participation in these elemental powers – their emplaced transcendence as they become realized in the native occurring of things in nature – involves us in the cosmos and provides a kind of intimate knowledge of it that far outruns what knowledge of the laws of science can supply. Such powers condition and are expressed in all being, occurring, and hence provide something that is common to all. They constitute shared points of reference, touchstones, that can act so as to offer a sense of community for all and hence a source of mutuality and understanding. If there is to be such a thing as a non-anthropocentric environmental ethic, surely, such a source of mutuality would be a relevant consideration in terms of how it would be grounded. And harking back to the idea to which brief allusion was made previously, an understanding grounded in participation in these elemental powers would seem to be a necessary condition of interpreting any messages that nature might voice.

To summarize, discussion of the two phenomenological vignettes given previously has revealed a number of key aspects of the natural world. First, it draws attention to a fundamental way of engaging with it that celebrates experience of its immediate occurring. This is taken ontologically to undercut, and to stand in contrast with, abstract discursive scientific descriptions that turn spontaneous things with their ever-changing countenances into static definable objects (including when one of their defined properties is that of change). Second, it reveals that this occurring of things in nature is always emplaced and that a mutual sustaining pervades their standing forth such that each is conditioned by the presence of its neighbours – all others that share this place. Third, there is an important sense in which things in nature befall us as ineluctably other, non-artefactual and epistemologically mysterious, such that even those things in nature with which we might be most familiar retain an element of strangeness and can be a source of wonder. They possess an integrity that we can sense but that we can never fully penetrate, comprehensively understand. Fourth, in sensing this integrity, we become aware that things in nature possess *intrinsic* value. They have lives and purpose of their own and can radiate a kind of normativity; their existence is not to be conceived exclusively in terms of how they might serve us. They do not exist simply as objects that are at our disposal.

Other aspects of nature's ontology will arise in later discussion. And the aspects identified previously will receive further elaboration and refinement during the course of this book. But for the present it might be useful to pause and to consider in a preliminary way some objections to which the account given so far might be thought to be vulnerable. They take their start from reservations about an unwarranted romanticism that it might be taken to express.

The charge of romanticism

Is the account of nature already given overly romantic? For example, it might be said that the phenomenological illustrations that I offer are skewed in their portrayal of nature, revealing only what might be considered to be its pleasing, "gentle", and harmless side. What of nature as "red in tooth and claw"? What would be the effect on the argument of including a vignette of a lion taking down an antelope, or a forest fire resulting from a lightning strike, or the desolation caused by a volcanic eruption or a tsunami? In addition, it might be said that the account lays too much emphasis on being receptive to nature. Where, in the picture that I have presented, is there recognition of the need sometimes to *oppose* nature, to combat disease and pestilence – for example, to prevent the spread of Ebola? Is there not a risk of underwriting a false aestheticism and a quietism that only those comfortably situated in the West for the most part can (at present) afford to indulge?

A second objection takes the reservations expressed in the first to a more general level. It claims that the view expressed seems to be based on a naïve phenomenology whose revelations ignore the tacit interpretation that is occurring throughout, and the unacknowledged narratives that inform it. Furthermore, it might be claimed that in addition to the account being misleading and theoretically incomplete in this respect, if these narratives were to be made explicit, they would be seen to express a range of highly questionable motives and prejudices that include the romanticism already noted. Regarding this latter – and with the likes of Rousseau and Wordsworth in mind – Andrew Stables (2009) claimed that the Romantics critiqued and problematized nurture while worshipping and mystifying nature. He went on to say:

> Romantic education, at least in its modern legacy, is thus both humanistic and anti-humanistic, simultaneously valuing the individual while devaluing the network of human relationships we call society, either through urging solitude (in a tradition that can be traced from Wordsworth to the later American transcendentalists such as Emerson and Thoreau) or through the creation of alternative, pre-lapsarian communities (as attempted, unsuccessfully, by, among others, Wordsworth's collaborator, Coleridge).

Concerns of the kind that my account draws on, or implies, romanticism and that this by its very nature mystifies and confuses in a way that veils questionable motives are likely to arise periodically throughout this work and there will be a need to respond to them on the particular occasions on which they again become relevant. With respect to the present context, I offer the following responses.

Certainly, the two vignettes were chosen to bring out positive features of the natural world and our participation in it. They are not intended to embody all aspects of nature, rather they are offered as foregrounding particular aspects that are occluded by some highly instrumental ways of engaging with nature that are taken to be prevalent in much everyday life, and with which they can be contrasted. Their function is to provide "simple" examples of experiences that are readily accessible (perhaps, for some, waiting on their doorstep, as it were) and yet that can illustrate highly significant facets of nature's ontology – its way of being. But as earlier discussion of those aspects of the natural world that involve pain and predation show, the harsh side of nature is not denied. Nor does it conflict with the mutual sustaining involved in the occurring of things previously described, for this is an *ontological*, not a *biological*, sustaining. The being of, say, the power of the predator and the vulnerability of the prey are mutually sustaining – are integral to their occurring. In participating in the mutuality of this relationship, we are opened up to a realm of agency that is in part constitutive of them, and also to a deeper sensing of the occurring of each – predator and prey – in the nobility of being themselves as the biological creatures that they are.

If an invitation to step beyond what is now routine experience of things so as to glimpse them in ways that reveal qualities that are "higher" or "deeper" than, say, instrumental or economic perspectives allow is deemed irretrievably romantic, then the account is indeed romantic. But then, no apology need be made for it. Too often, the term "romantic" is applied as a term of derision – as if the kind of experience to which it refers somehow misleads us concerning the nature of reality, substituting an unwarranted sentimentality for what "really" is hard, unforgiving, materialist, and commercial – perhaps, ruled by iron laws that brook no deviance and that cast alternative portrayals as frothy fiction. Indeed, so toxic has a romantic view become for some that a recent paper on environmental issues explicitly avoided using the term "nature" altogether partly because it was held to have Romantic associations (Griffith and Murray, 2017). While romanticism can be prone to becoming overblown, it is far from self-evident that, properly understood, it does not contribute a vital dimension to our experience – one that possesses its own integrity and truth. Certainly, as a cultural/philosophical movement, it has been upheld by many as making an important positive contribution to the evolution of Western sensibility.

In his *Roots of Romanticism*, Isiah Berlin (2000a) suggests that while Romanticism is very difficult, if not impossible, to define, it is possible to identify two themes that lie at the heart of the Romantic movement. The first of these is that of man's (*sic*) indomitable will – including the idea that men do not discover values, but create them. This line of thought burgeoned in the late nineteenth and early twentieth century, reaching its apogee in Sartre's existentialism where the individual's ultimate freedom to define itself through its choices is acclaimed. The second theme identified by Berlin is that there is no fixed underlying structure to reality: nothing can be (even in principle) finally and fully known. Because everything is in constant motion, there is only the endless self-creating of the universe that man can either join in, throw himself into, or can attempt to oppose by trying to nail things down and give definitive rational accounts of. For the Romantic, this latter is futile – it empties, deadens, what it purports to describe. Full accurate description is impossible – hence, the importance of allegory and myth in communicating that which can never be fully expressed. There is no set pattern or underlying system of reality. As Berlin puts it, there is no "jigsaw" to be completed.

If, for the purposes of the present argument, we accept this analysis, clearly, with regard to the first of the themes identified, there exists a considerable tension with the account that I give. While, in later chapters, I will be discussing the senses in which we *participate* in the being of things in nature, certainly this will fall short of humankind being set up as either the sole author of values or the sole author of itself. In this sense, my account is not wholeheartedly romantic in Berlin's terms. With regard to the second theme, however, there clearly exist some strong resonances. Other commentators (e.g. Marshall, 1995; Garrard, 1998) have noted the valuable contribution that Romanticism makes to our understanding of, and relationship with, nature. For example, they draw attention to the ways in which it encourages a positive, open, and receptive relationship with the natural world – a relationship that through the participation of the imagination and the affective capacities of human beings facilitates an active simpatico with it. This sense of a kind of underlying union with nature is the reverse of the attitude promoted by the scientistic thinking that was sired by the Enlightenment, and it opens the door to a fuller and more wholehearted engagement. Surely, for the West, this constitutes a hugely important cultural resource when it comes to shaping a more adequate orientation towards the natural environment than that which much of our current behaviour expresses. And while it is partial, it seems to me that the sentiments expressed, for example, in Wordsworth's *Tintern Abbey* (amongst many of his poems) express an important and valid potential dimension of our experience of nature. His evocation of the power, beauty, and steadfastness of nature and its ability to affect us so as to be "sensations sweet, felt in the blood, and felt along the heart . . . passing even into my purer mind, with tranquil restoration" are surely recognizable to many who take the opportunity to attend to the natural world.

Notwithstanding Dr Johnson's dismissal of nature as displayed in Scottish scenery as a wide extent of hopeless sterility, experiences of which do nothing either for the imagination or the understanding, surely the poetry of Wordsworth, Keats, Hopkins, and Clare offers possibilities of an enlarged sensibility that is both valid and significant. And certainly this sensibility is not one of an overweening rhapsodic optimism. *Tintern Abbey* itself is far from being this. Rather, it intimates a more sombre side when it speaks of "the heavy and the weary weight of all this unintelligible world" and "the still sad music of humanity". It is true that, for example, Keats' *To Autumn* portrays nature exclusively as harmonious and fruitful – shot through with aesthetic and material bounty that is both a comfort and inspiration to humanity. It might be said that this is a very partial and "romanticized" view of nature that omits the equal reality of pestilence and famine, and the human effort involved in selecting and organizing nature's potential (the poem's apple-bearing "moss'd cottage-trees" do not occur there entirely of themselves). It is understandable that some might find this portrayal as too cosy by far. But, nonetheless, Keats' poem does alert us to something that, too, is real and that is easily overlooked when we are preoccupied with everyday busyness or self-oriented anxiety. To be sure, the view is partial, but it has validity as a response to a particular field of "occurrings" that are a legitimate part of a greater whole.

Returning to *Tintern Abbey*, it is easy to detect references to aspects of the human condition that are not usually considered to be matters for celebration. Romanticism's acute sensitivity to the passing of things of great value, and feelings of lingering sadness evoked by a sense of current or potential loss, are together one of its most compelling features. Much of this seems to arise from a recognition of human finitude/mortality/frailty, but it also arises in response to perceptions of the consequences of human arrogance, avarice, and associated lack of receptiveness and imagination. (Here one might recall, for example, Wordsworth's "Getting and spending, we lay waste our powers".) It seems to me that the work of latter-day romantics such as Housman and Tolkien are still resonant with all this. In the case of *Tintern Abbey*, the latter part of the poem alludes to such things. Here, Wordsworth anticipates a parting from his sister (and conceivably his own demise?) when their shared ecstatic experiences of nature live on in her and can serve as a bulwark against fear and pain. But she in turn will die, whereupon the sheer immediacy of those momentous moments will be utterly extinguished – that they ever occurred remaining only as a distant echo in a poem. Preoccupation with mastery occludes, or deeply subverts, all such sensibility to the human condition – perhaps, in part represents an escape from it?

Regarding the more general criticism that the whole account is founded on a naïve assumption that somehow phenomenological descriptions give access to some pristine world of things as they are in themselves, several points may be made by way of rejoinder. The first thing to say is that it is true that all perception and understanding involves interpretation. Things are always experienced from some perspective or other – from within a particular horizon of significance. As Martin Heidegger notes, from the very beginning of our relating to the world a "fore-structure" of understanding is in play that is our "first cut" of meaning, enabling an entry into an intelligible world. It provides us with meanings that ground further interpretations (Heidegger, 1962, pp. 182–195) and that can themselves come up for reinterpretation. In this sense, all human understanding does, indeed, occur within the "hermeneutic circle" of which Heidegger (and others) speak. But it is important here to recognize that that we never start from nowhere, or from some purely objective, neutral position, is not a defect of understanding, but a *condition* of it. A similar point is made by saying that all experience is conditioned by our form of sensibility that supplies an initial stance towards, and framing of, everything that we encounter. (I will say more on this presently.) Now, if, on the back of this underlying truth, it is being claimed that as all phenomenological descriptions reflect antecedent interpretations this vitiates any possibility of any of them being primordial, the question arises as to whether there is any basis on which to privilege one such description over another. For example, returning to the description of the woodland dell given previously, why privilege this over the experience of say a logger whose overriding experience of the beech tree is in terms of potential board-length and market prices? Or, as aforementioned, that it is redolent with an indulgent aestheticism? The response to this kind of charge has a number of components.

First, there is an important sense in which some human experiences of the natural world possess a particular fundamentality. In addition to being immediate and concrete, they will be those experiences that are (a) capable of being widely had; (b) most open to the manifoldness of nature; (c) capable of involving all the senses; (d) experienced as being of the things themselves – i.e., are experienced as involving a "responding". Of course, these "criteria" shade into one another and are loose in the sense that they can admit of a wide range of different and, perhaps, sometimes conflicting experiences. I offer them as indicators or pointers rather than as definitive categories. But they are enough to guide us in evaluating experiences of nature in a general way. For example, narrowly instrumental perceptions would be excluded under "b", averaged off everyday perceptions under "c" and "d", deeply mystical and esoteric under "a", and so forth.

Second, it is experiences of this kind that form the ground of intelligibility for abstract notions, such as those produced by science. This is to say that ultimately the intelligibility of such notions as "gravity" or "mass" is conditioned by the extent to which they can be translated into the world that these experiences open up – a world where things are immediately and sensuously present, and can be felt – perhaps, handled. This point will be elaborated in the next chapter when ideas of embodied consciousness and understanding will be broached.

Third, and closely connected to the previous point, the language in which other experience is articulated will be heavily derivative on the language through which primordial experience is expressed. (Incidentally, by this locution I do not mean necessarily to imply that the experience existed independent of language.) This is not to say that through various feedback loops, the "original" language will not become modified and enriched through an absorption of, say, language that arose in more abstract contexts. But there is an important sense in which the language that has evolved through our encounters with "middle-sized solid objects" (historically, many of them natural things), and the understandings and emotions that they evoke, will be prominent in the horizon of significance against which our meanings and interpretations arise. David Abram (1997, p. 34) makes (perhaps extends?) the point nicely:

> Our spontaneous experience of the world, charged with subjective, emotional, and intuitive content, remains the vital and dark ground of all our objectivity.

Finally, it must be made clear – if it is not already – that by primordial experience here, I do not mean experience of some putatively neutral and pure kind such as that once claimed by sense-data theory that was prominent in analytic philosophy in the first half of the twentieth century. Rather, I think that there are some resonances between the view that I am propounding and the position of John McDowell (1996) on the re-enchantment of nature – both in terms of the kind of thing nature has to be in order to constitute a source of validation of empirical beliefs and by implication what would be involved in knowing nature. Because it represents a way of restating and reinforcing my position that draws on thinking developed in a different philosophical tradition, and hence illustrates some parallels, it will be helpful to consider some aspects of his view on the matter.

A key point in McDowell's account is a rejection of equating nature with the logical space in which science locates it – a space of blind universal laws. He argues that if experience were to be construed as made up of impressions – "impingements by the world on a possessor of sensory capacities" – and this world is the world described by the natural sciences – i.e. this is the logical space in which they function – it is *different in kind* from the world of normative relations that constitute the logical space of reasons (where, for example, there is talk of one thing's being warranted, or correct, in the light of another). And if, in this way, the logical space of reasons is *sui generis* compared with the logical space allotted to nature, then it becomes impossible for experience, as construed previously, to act as the tribunal for empirical *thinking* because they operate in accordance with completely different logics and therefore cannot, as it were, interrelate. They are simply incommensurable. This means that the idea of empirical thinking itself becomes incoherent. Hence, McDowell argues that reason in the sense of the operation of concepts goes all the way down to the level of our most primordial experience of things. There is no prior apprehension of a pre-conceptual "given", upon which concepts then operate. How could they obtain a purchase or participate in an alien logical space? Concepts are involved in our apprehension of nature from the bottom-up and nature is therefore no longer necessarily divested of everything normative.

Now, it might seem that this account runs counter to my emphasis on the otherness of nature and its foundational role as a given. It is true that McDowell wishes to repudiate what he terms the "myth of the given". But here it is important to be clear about what is meant by "the given". If we mean by the given some sort of absolutely pristine sensory experience – such as was, indeed, postulated by the "sense data theory" of, for example, G. E. Moore (1953), H. H. Price (1932), and A. J. Ayer (1940) – there is no more room for it in my account than in McDowell's. As I have previously emphasized, we are always in the world understandingly and even that favourite example of sense data theorists of the experience of isolated after-images of coloured patches, these are always understood as occurring in a certain context and possessing a particular significance. To borrow a term taken from a different philosophical tradition, they – and all experience – are always had within some prior meaning-giving horizon of understanding. But this does not make the being of nature transparent, deny its inherent mystery in its self-arising. Quite the reverse. Mystery is only possible in the logical space where significances are in play. Mystery, too, is just such and so.

Of course, this still leaves room for debate as to how developed, abstract, reifying, and systematized the concepts that inform primordial experience need to be. It will be argued in the chapters that follow that the degree of reification and discursiveness involved will be pretty minimal – such that some might baulk at referring to them as concepts in any full blown sense. But the central point that experiences of nature occur within, as I would put it, a form of sensibility that is shot through and through with human significances and is in that limited sense rational and a cultural product has to be granted. Only on this premise can nature be apprehended at all, and so construed it is capable of possessing a rich founding and normative dimension. It can be rightly conceived now as a "given" in the sense that it is both not simply a product of our decision-making and choice – as McDowell (1996, pp. 10–13) concedes, there is an essential element of human passivity in our perception of it – and that as an orientating idea, it is so deeply embedded in our form of sensibility that it is constitutive of our way of seeing and understanding the world both cognitively and affectively. Nature is not, for us, a disposable idea. Furthermore, by extricating it from the logical space in which conventional science locates it, it is made clear that the kind of unity and transcendence that nature possesses fundamentally need not be that of the highly abstract laws and closely defined and categorized objects in terms of which the natural sciences characterize it. Some important curriculum implications of this argument will be explored in Chapters 6 and 7.

Notions of development and the issue of sustainability

Problems regarding the idea of sustainable development were flagged in Chapter 1. The phenomenology of nature that has been elaborated in this chapter allows us to refine some elements of the criticisms set out there. They revolve around the relationship between ideas of "sustaining" and "development" after it has been granted that nature possesses its own integrity – as argued throughout this chapter. Once this is accepted, and with regard to thinking about the natural world, "sustaining" takes on strong conservationist meanings that embody ideas of respecting and nurturing things as they are in their own nature. In contrast to this, "development" as understood in the context of Brundtland-type definitions of sustainable development, emphasizes the idea of humanly planned change: an anthropocentric ordering of things so as to produce more or "better" than what currently is available. As noted in Chapter 1, in practice and under the influence of Western style culture, such development is construed in terms of ongoing materialistic economic growth that in turn requires ever more extensive manipulation and exploitation of the natural world.

To cast the idea of development in these terms is not of course to deny that development is inherent in the natural world. Quite the contrary, it is rather to recognize that the kind of development involved in dominant interpretations of sustainable development is of a radically different sort to that found in nature (Bonnett, 2017a). As a prelude to elucidating an authentic view of sustainability, it will be helpful to elaborate on this. Fundamentally, the difference between these two kinds of development can be characterized in the following way: from the perspective of things in nature, economic development proceeds in accordance with an "external" norm. This is to say that its goals bear little or no relationship to the "internal" norm of the thing's own development – that is, what would count as an ongoing expression of its own integrity. An example of this would be commercial plant and animal breeding programmes that seek to "improve" upon nature in the sense of engineering outputs that better meet current human requirements, but that no longer relate to the thing's flourishing in a natural environment. Extreme (but common) cases of this would be where a fruit has been bred to increase its sweetness, succulence, and "peelability" while contributing nothing towards its fertility, or where an animal is bred to raise the bulk, protein level and tenderness of its flesh, such that it would no longer survive in the environment in which its predecessors thrived.

To draw out this distinction between internal and external norms of development is not, in itself, to censure the domestication of plants and animals. Humanity as a whole can hardly return to the state of hunter-gatherer. Rather, it is to establish a distinction that later will be held to be relevant to the idea of living in nature that was alluded to in the previous chapter. Also, it enjoins us to be honest about the character and extent of the denaturalization that is being enacted under the aegis of some versions of sustainable development and to recognize the underlying anthropocentrism that energizes it. This recognition of denaturalization can give a perspective on other issues sometimes associated with establishing sustainability, for example, the advocacy of genetically engineered crops to meet the future food needs of a growing global population. Here, at what now appears to be an even more profound level, nature's integrity is willingly to be sacrificed in the service of what are taken to be human interests. (The question of human population growth and some of the normative and ethical issues that it raises will be discussed in Chapter 5.)

In the meantime, and contrary to this inclination, it is argued that at the heart of sustainability lies the idea of allowing things in nature to be themselves, rather than to re-engineer them in accordance with criteria that do not emanate from themselves. When this latter occurs, things that were once natural become detached from nature, they are no longer spontaneously vitalized by its powers, processes, and rhythms, no longer an integral part of, say, the play of the seasons, but rather dependent on the playing out of human purposes that increasingly show a disdain for what is given by nature. In addition, as previously noted, given the complexity and temporal and geographical extent of natural systems, there is little reason to suppose that these purposes will be anchored in an adequate knowledge base. This makes the growing ambition of humans to manage large-scale natural systems such as "the oceans" a frightening prospect. Philosophically, it leads to an examination of ideas of human stewardship of the Earth that will be undertaken in the final chapter.

For the moment, it is perhaps appropriate to flag what looks like a fundamental choice that lies before us – a choice between two paths. The first path consists in the ongoing pursuit of human dominance over the natural world, and a consequent evermore extensive denaturalization of our environment. This is happening at a number of levels, for example, in terms of not just the physical environment that we inhabit, but also, and more fundamentally, the psychological and intellectual environment that conditions our thoughts. The second path consists in an attempt to be guided by and to live within and alongside nature's internal norms insofar as this is compatible with living a healthy life. (It might also be part-definitive of what would *count* as living a healthy life.) Together, consideration of these alternatives raises a further question that goes to the kernel of all environmental debate and that implicitly or explicitly conditions all environmental policy: what are we entitled to expect from nature, that is to say, what are we *warranted* in requiring from it? Under the sway of a metaphysics of mastery and an implicit anthropocentrism, these hardly arise as issues, but once brought to the surface they evoke a range of complex considerations, which is a central task of this book to unfold.

To anticipate one of these – and by way of illustration – in the chapter that follows I will discuss the phenomenon in Western societies of an accelerating trend of glorifying the artefactual and a corresponding disdain for the given. Here, it will be argued that, increasingly, all aspects of our environment are being brought under our control and re-engineered to better fit whatever our current purposes and desires happen to be. And not just our environment: for some, in many respects, our own anatomy, too, has become a matter of choice as techniques are developed for refashioning it in myriad ways. What is simply given is no longer good enough. Why should we be content with it? Under the aegis of the metaphysics of mastery, it appears to be almost self-evident that we should impose our will on all that matters to us. At best what has been given is to be regarded as simply an initial resource, so much brute material to be reworked, improved upon, brought into line with what will satisfy our desires.

As I have said, the philosophical roots and implications of this will be developed in the chapter that follows, but before moving on to that, it will be helpful to relate this cultural attitude back to the current discussion concerning internal and external norms of development. For this discussion contains the seeds of a somewhat different approach to life and to nature. It also opens up a key aspect of what it is for humankind to be *in* nature rather than putative overlord. Put simply for the moment, if we are attentive to nature's internal norms – if we are able to gain a sense of what these are – and if we allow them to play a significant role in guiding our thinking and behaviour – then we allow nature to flow through us. In this sense, we truly become part of nature. Human consciousness becomes nature's consciousness. Such an argument raises many complex issues concerning how we are to discern nature's internal norms and precisely why we should accord them any special status as guides to right thinking and behaviour. Certainly, there are dangers here. These issues will be developed in Chapters 4 and 6. For the present, I simply raise the prospect of a frame of mind (and body) in which to a significant degree human–nature and culture–nature dualities might be dissolved and in which the given is celebrated as a gift that initializes life rather than obstructs it.

Note

1 I first developed this idea in *Retrieving Nature*, where I drew on Martin Heidegger's interpretation of the ancient Greek term "phusis".

3

TRANSCENDENT NATURE AND ITS ENEMIES

Chapter 2 elucidated a phenomenology of nature that reveals it as holistic, transcendent, and "other". It distinguished this from the conception of nature offered by scientific ecology in which there is a sense in which these features are maintained, but they are understood in a different and abstract way. Here nature's holism is portrayed in terms of definable entities embedded in a web of causal or probabilistic law-governed biophysical interdependencies. In contrast, the phenomenological perspective focuses on the being of natural things in our experience – the characteristics of their native occurring in our immediate encounters with them. This is taken to constitute a primordial experience of nature, one in which natural things are disclosed as themselves in their transcendence rather than being more exclusively the product of the categories and theories that we project. As themselves and in their being or occurring – that is to say, *ontologically* – things in nature were described as displaying a number of key qualities. For example, in their standing forth, their holism is revealed as thoroughly organic in the sense that it consists in a fluid mutual sustaining such that the standing forth of any individual thing resonates with that of its neighbours, and both contributes to and is conditioned by the ambiance of the place in which it occurs. The occurring of each is imbued with the occurring of all else that is present in that place. As embodied beings, we can participate in this native occurring of things, but we do not determine it. Indeed, it was held that in many respects, it is redolent with inherent mystery. Yet through receptive participation over a sufficient period of time, we can become aware of the inherent integrity of a natural place and intuit a sense of what counts both as its well-being and the well-being of its constituents. With this, we intuit what responses on our part would be fitting and what would not. Here intimations of nature's normativity and intrinsic value are given, as are expressions of elemental powers such as birth and death, appearing and withdrawing that pervade the occurring of all natural phenomena. These, then, are some key characteristics of transcendent nature from a phenomenological perspective.

Clearly, to the extent that these characteristics are accepted, they have highly significant implications for how we should think of and relate to the natural environment – and hence the basis from which we should address the prospect of environmental catastrophe that now confronts us. But the whole idea of nature as some transcendent other, and the further idea that somehow it has inherent normativity that we can intuit and to which we are ethically challenged to respond has been seen to be problematic and has many detractors. Objections take a variety of forms and arise from a variety of sources. In order to establish a foundation for subsequent argument, it will be helpful here to address some of the main lines of resistance to the idea of nature's transcendent reality.

Criticisms of the reality of nature and its contemporary relevance

There are many senses of the idea of "nature" and of what is "natural", and many criticisms concerning their integrity. These include their vagueness, their lack of cultural or historical stability, their capacity to incorporate implicit and questionable values and power relations, their lack of intelligibility and/or relevance in our late-modern historical period, and their potential for endorsing what some see to be a false dualism of nature and culture and for homogenizing the former. Indeed, in a general way, is it not the case that, as Marshall Sahlins (1977, p. 105) once observed, there is a propensity for human societies to invent the nature they desire or need, and then use it to justify the social pattern that they have developed? Here we have a perpetual movement back and forth between the culturalization of nature and the naturalization of culture that belies the idea that there is any such thing as transcendent nature. Undoubtedly, these criticisms contain important elements of truth and they raise many complex issues. As I have addressed many of these elsewhere (see, for example, Bonnett, 2004) and they are not all germane to the current argument, in what follows I focus on responding to a set of views that query the reality of nature and its relevance in the late-modern period.

One such view argues that because there is now no part of the biosphere that has not been affected by human activity, nature conceived as some autonomous or self-arising realm has lost its meaning – it is "at an end" and we live in a post-natural world (McKibben, 1989). Alternatively, Anthony Giddens (1994) claims that in our post-traditionalist culture, nature as formerly understood has become "socialized", has "dwindled away" such that it "no longer exists". Along with tradition in general, nature's authority as a given "external" context of life has evaporated. This is held to be true in at least two senses. First, we can – and increasingly do – question what has traditionally been held to be "natural" or "not natural" as a guide to how we should behave. For example, many traditional norms governing the behaviour and social roles of children, women, and men based upon what was taken to be consistent with their "nature" have been heavily modified or abandoned altogether during the late-modern period. Second, modern science and technology make it possible for us to modify many of what formerly would have been accepted as natural givens, such as crop yields, the progress of disease, and our own anatomy. We now have so much more power over things of this kind that in many respects they become matters of human choice. The upshot of arguments of this sort is that nature conceived as some kind of reality external to our personal intentions that either does, or should, determine our lives now exists only as a chimerical human reinvention – an idea whose day has passed and that sometimes we might resurrect artificially in an attempt to make play on the echoes of its former influence (as, for example, in advertising food and medicinal products) or out of nostalgia. This view resonates strongly with postmodernist philosophers such as Richard Rorty (1980) who have seen nature as a concept that was the product of an optional humanly constructed narrative that is now defunct.

Implicit in these criticisms is an underlying theme that severely undermines the authority of nature at the level of principle: essentially nature, as we have it today, is (merely) a social construction and as such is susceptible to reflecting differing cultural perspectives, historical periods, and the interests of social elites. Hence, as an idea, it lacks the solidity and impartiality necessary for guiding an understanding of human flourishing or for founding an environmental ethic. The situation within an influential field of academia has been summed up by Ursula Heise (2008) in the following way:

> More broadly, the basic goal of work in cultural studies for the last twenty years has been to analyze, and in most cases, to dismantle appeals to "the natural" or "the biological" by showing their groundedness in cultural practices rather than facts of nature. The thrust of this work, therefore, invariably leads to skepticism about the possibility of returning to nature as such, or of the possibility of places defined in terms of their natural characteristics that humans should relate to.
>
> *(p. 46)*

This sceptical stance is reinforced by other arguments that call into question the very senses through which customarily we take ourselves most immediately to apprehend nature. For example, Ulrich Beck (1992) has noted how a report of radioactive leakage can shatter a sense of normality and tacit everyday embeddedness in nature hundreds of miles away by affecting our ways of perceiving and relating to the most familiar of things, such as the air that we breathe and the water that we drink. In this sense, we can both become detached from our feelings of being securely emplaced in nature and have our faith in the senses that emplace us undermined because the latter cannot perceive even highly dangerous levels of toxins such as radioactivity. In contrast with the view of Richard Louv cited in Chapter 1, that heralds the "re-naturing" of everyday life, Beck foresees that it will be communities formed by a sense of transgressive risk rather than an affiliation to nature that increasingly will define modernity.

The upshot of these lines of thought is a degrading of appeals to any direct sensing of nature in favour of a legitimation of abstract discursive sociocultural and scientific accounts that are taken to provide more sophisticated portrayals of how things really stand by making the understandings embedded in direct experience suspect and subservient to theoretical fabrication. Indeed, when it comes to revealing the fragility of nature on a global scale, it has been held that we need precisely to disengage from immediate experience of emplaced nature and to attend to the products of complex computer modelling – "third nature" (Wark, 1994). Here, in effect and from an ecological point of view, we are presented with three ideas of nature: "*first* nature" as that which exists entirely innocent of the effects of any human activity; "*second* nature" as that whose existence reflects the presence of human interventions; and "*third* nature" as that which exists only through the medium of abstract human theorizing and modelling activities. And on the view being described, in its various iterations, first nature now is either defunct or irrelevant.

Here, then, we are confronted with some serious and influential criticisms of allowing the idea of nature any prominent place in environmental discussion. Yet it seems to me that this is not necessarily the right way in which to respond to them. Rather, such criticisms can be seen to be helpful in elucidating a conception of nature that remains substantive and valid, and that properly understood demands a central role in such discussion. For example, McKibben's claim of the "end" of nature alerts us to the error of equating nature with wilderness – some pristine realm of physical entities totally untouched by, and unaffected by, human activity. It is probably true that this no longer exists (on the Earth), but it seems to me that to grant this is in no way to deny the validity and power of the concept of nature. Rather, it can lead us to recognize a sense of nature not as any such realm, but as a *dimension of experienced reality*. As a dimension of experience, it can be present in varying degree, to wit in the toughness of a steel artefact as well as, say, in the ambience of an ancient forest. The underlying point is that in both cases, there comes a point at which we come up against that which is ineluctably other, not human-authored. Although the steel is wrought by the human hand, its particular chemical composition and properties designed for a purpose, ultimately we have had to work with what has been given in the first place by nature. This partly determines what is possible and what we experience. However far, say, the science of metallurgy goes in extending the frontiers of what can be produced, there is always a frontier – a prior encounter with nature (Bonnett, 2015a). Certainly, steel is an artefact, not a natural product, but ores that are the result of natural processes play a profound role in the genesis of its existence and its properties, and what we understand it to be.

Take now the criticism that nature is essentially a mutable human construction. While undoubtedly concepts of nature are socially produced in the sense of arising in the context of a form of human sensibility, central to the meaning and experience of nature *qua* nature is that precisely it is *not* socially produced. Rather, as discussed in the previous chapter, we experience it as essentially non-artefactual, something that exists completely independent of human intention, occurring or arising out of itself. It was argued that in this sense quintessentially nature is *self-arising*. It is true, of course, that we articulate our experience of, say, a wild flower underfoot or some distant constellation faint in the night sky by means of concepts, and concepts, indeed, are socially produced. In this sense, they are "constructed" artefacts. But *that* these things exist in those moments of experience is not simply down to us and our concepts. Although, as will be argued more fully in the next chapter, there is an important sense in which they occur in the space of human consciousness, the fact of their sheer existence is theirs. Indeed, granting that concepts themselves are socially produced is not to say that they are entirely anthropogenic. Some, at least, arise in the process of an intercourse with a reality that is not of our making and in that respect are conditioned and shaped by enduring aspects of what is other. Hence, in these cases, we are not sole author of our concepts, they are not matters purely of our choice or decision, and the fact that we played a part in their genesis neither means that we can change them at will nor that we have exhaustive knowledge of what they mean.

This is nicely illustrated by the concept of nature. Contra the view of Rorty alluded to previously, the concept of nature as the self-arising other is so deeply embedded in our form of sensibility that it can hardly be conceived as "optional" (Bonnett, 2004, pp. 57–60; see also: Rolston III, 1997a; Crist, 2008). Without the sense of encounter with what is essentially other, the deep structure of our form of sensibility articulated in such ideas as those of reality and truth – and hence ideas of perception and knowledge – would collapse. These latter notions presuppose some reality beyond us to *be* perceived, described, known, etc., if they are not simply to implode – and with them, human consciousness. In addition, far from being an idea that has little cultural or historical stability, the understanding of nature as occurring from out of itself would appear to go back as least as far as the ancient Greeks (see Heidegger, 1975, p. 42, 1977, p. 10). Furthermore, it seems entirely plausible to claim that relating to the otherness of nature has shaped our senses over the millennia – indeed, fulfilling this function has largely (but not exclusively) been their *raison d'etre*. Yet nature remains mysterious to us in so many ways: we can neither fully comprehend it in its many countenances and entirety, nor comprehensively predict what encounters with it will bring.

It seems to me that observations of the foregoing kind are sufficient to defending the idea of a transcendent autonomous nature from the criticisms that were raised at the beginning of this chapter. But it is important to note that to claim a self-arising aspect as essential to our understanding of nature is not to deny that frequently – perhaps for the most part – how nature reveals itself reflects the purposes that we pursue with regard to it and the concepts that we employ to articulate it. For example, a mountain will appear differently to us depending on whether we are engaged in mining it for some ore or representing it artistically.

But these purposes and concepts can never fully determine what lies before us. Indeed, on occasion our ongoing application of purposes and concepts to our experience of nature can be rebutted: for example, our attention can suddenly be commanded by something quite other with respect to our current preoccupations. Perhaps, we are brought up short in a meditative river bank stroll by the momentary glimpse of luminescent blue of a darting kingfisher – or the high energy hum of a swarm of bees on the move. On these occasions it is nature that is the chief agent of our perceptions.

More generally, there is a complex and intimate interplay between cultural motives and artefacts and the otherness of nature that constitutes world-formation. For example, the design of an implement such as a spade both is shaped by the soil it cleaves and in cleaving brings to presence the resistance, texture, odour and shy lustre of the clay. The nature that we experience is human-related in that human consciousness (including, frequently, human activity) provides the place and occasion for its appearances, but not human-centred: we are neither the author of the living presence of natural things, nor do we assume that they presence exclusively for us. As previously discussed, we understand natural things as having their own lives and interactions regardless of whether we witness this or speculate it.

In sum, as quintessentially other-than-human, things in nature have aspects that always lie beyond us, withdrawn, as yet (and perhaps never) to be revealed. They have their own histories and futures, profiles and countenances that we will never see. This remains true no matter how developed our scientific understanding becomes. Indeed, preoccupation with scientific observation, categorizing, and explanation can lead to an attenuation of a more direct and intimate sense of the being of things in nature. For example, the quiet mystery of the sheer existence of some wayside flower and its subtle changes of hue and aspect in the play of sunlight can easily become occluded as we turn to identifying it according to generic characters listed in a flora database. When experienced in their native occurring, things in nature are epistemologically mysterious and retain the ability to offer invitations to participate in their being in unique and never wholly predictable ways.

Because of the extent and depth of their influence, I now return to two particular and trenchant enemies of nature so conceived and that were alluded to in Chapter 1.

Scientism and the metaphysics of mastery

In Chapter 1, I characterized scientism as the generalization of the authority of the methods, assumptions, and constructions of the classical experimental sciences beyond the discipline of science and into our daily lives. In this way, scientific approaches are taken to have some privileged access to reality as a whole – to reveal what is "really" real – and are accounted the arbiter of good thinking in general. I was careful to distinguish this set of transgressive presumptions from the activity of science, itself. With regard to the natural world, I also exemplified scientism as portraying ultimate physical reality in terms of atomic particles rather than familiar solid objects and as reducing complex animal behaviour such as the home-building activity of an animal like the beaver to the working out of blind quasi-mechanical processes. No doubt, these particular examples are highly simplified and do not represent the full range of conceptions and models of reality that experimental science currently employs. But it is not – and could not be – the intention of this necessarily brief characterization to cover all the possibilities here. The underlying point that these examples are taken to illustrate is that classical experimental science sets the natural world up as essentially mechanical, specifiable, and revealed to the disengaged observer. And these are the features of thinking that *scientism* promotes in the world at large.

Clearly, the position that I am developing challenges this presumption that somehow such scientific accounts are truer, more objective in the sense of providing a more authentic depiction of the world – one that properly reflects how it really is. Why privilege blind mechanical depictions of the natural world over those that speak of purpose and agency? Why be tempted to posit as fundamental a world of, say, colourless, blindly hurrying particles when human experience of the natural world is so much richer than this and cannot be adequately articulated through its vocabulary? I will return to these questions shortly. For the moment, I make the point that while quasi-mechanical conceptions might be quite acceptable within the discipline of science with its particular project towards the world, and where its limitations as well as its strengths are recognized, given the richness of experience and depths of intelligibility that they deny, such reductionism looks highly arbitrary when it gets generalized, as with scientism. For example, not only do natural historians unremittingly describe the activities of many animals (and sometimes plants) in terms that ascribe agency to them – the lion *hunts* its prey, the bird *builds* its nest, the parent *protects* its young, the tree roots *seek out* water – it is far from clear that we could articulate or understand such phenomena in any other terms. The vocabulary of purpose and agency is indispensable in supplying a degree of intelligibility to these features of the world. As previously noted, we can never have insider knowledge of, say, animal behaviour, but in order to make some sense of it, we require the vocabulary of agency and purpose. And it is not obvious that we should regard such articulations as any more figurative than when employed to describe human activity.

From a phenomenological perspective, the shortcomings of the language of the physical sciences with regard to articulating the whole of nature go beyond its inability to do justice to the element of agency exhibited in animal behaviour. It also fails to recognize a further highly significant and closely related feature that is salient throughout our experience of the natural world: the general presence of *anticipation*. In the previous chapter, it was argued that in their occurring natural things participate in a place-making. It was also argued that in being *there*, we, too, participate in this place-making. It seems to me that a fuller recognition of the impoverishment of thinking that scientism brings, and a consequent loosening of its grip, can be achieved by giving further attention to the fact that we live in places and that this latter is not a contingent matter; we always and necessarily experience ourselves as emplaced. As exemplified in Chapter 2, places in this sense are not simply locations that, as it were, are slotted into some pre-existing space-time causal matrix. Rather, they occur as distinct neighbourhoods with their own particular ambiances and internal relationships between the things that populate and constitute them. They create their own space. Here, it was argued, a sense of non-human agency and purpose can become salient. Taking this further, phenomenologically, it can be seen that a focus on place foregrounds that the way that things participate in our world can be inherently anticipative. This can occur on myriad occasions. For example, we can experience the parched soil as awaiting the rain, or some gaunt weathered shoreline rock as anticipating the coming storm surge. Swelling buds standing out on dark stems anticipate the warmth of lengthening days, the deep-throated foxglove the visiting bee, the spider's web the stray fly.

Of course, under the influence of scientism, a rejoinder to this would be "Not really". Take the rock. This is an inanimate object made of inert material. It anticipates nothing. It has purely physical properties such as colour, hardness, solidity, and weight. Or again, some will say that more fundamentally it possesses mass rather than weight, reflects light of a particular wavelength rather than itself possessing colour. But why stop here? Why not say that really it is a collection of atoms or more fundamental particles, or perhaps an energy system? And so forth. In discussions about the nature of reality, this is a common progression. At this point, two key questions arise: (1) On what basis do we decide where to stop in this reductive series? (2) With what justification do more scientific descriptions (i.e. those of classical experimental science that still dominates popular debate and its underlying assumptions) trump those of other experiences when it comes to describing what is "really" real? Proximally and for the most part, we do not live in, say, a colourless world of hurrying atoms (or of some other supposedly purely objective realm of inert entities and mechanical forces), and we do not live in a world where things have mass as distinct from weight. Indeed, it is arguable that our grasp of mass is grounded in our experience of weight and that "mass" only has logical priority in the context of an abstract system. Looked at dispassionately, we might say that there is no *one* reality, none that is more fundamental than another in an objective sense. Yet, as noted in the previous chapter, ultimately, any account of reality has to be cashed out in terms of what humans can directly experience in order to be understood, and so features that are central to this lived reality do, indeed, possess a particular kind of fundamentality when it comes to characterizing the human situation and our place in the world.

Harking back now to the question raised earlier in this section concerning the privileging of reductive mechanical descriptions of reality, my argument is that the reason why, despite the foregoing considerations, this sort of account has gained the ascendance when it comes to portraying underlying reality is that it facilitates an ultimately overweening will to mastery that the other richer accounts do not.

Hence, I suggest that the answer lies in the holding sway of what previously I have termed the "metaphysics of mastery" and it is now time to elaborate this idea. In Chapter 1, I referred to the way in which dominant strands of Western culture increasingly frame all issues in terms that are deeply human-centred and manipulative, and that the underlying motive is that everything is to be understood not simply in terms of serving human flourishing (which it would be possible to conceive in receptive and celebratory terms) but in terms of the assertion and satisfaction of the human will. Now, amongst other things, what we have here in many areas of life is a pervasive disdain for given reality and the authority of nature and an accelerating preference for the artefactual – the products of our own ingenuity. Hannah Arendt (1998) observed that in late-modern times man (*sic*) seems:

> possessed by a rebellion against human existence as it has been given, a free gift from nowhere (secularly speaking) which he wishes to exchange, as it were, for something he has made himself.
>
> (*pp. 2–3*)

It is perfectly understandable that, as centres of reflective consciousness, we should value our own flourishing and that, for example, this is a necessary element of any plausible notion of sustainability. However, it remains an open question as to whether elevating our own flourishing above all other considerations (such as the survival of other species) is justifiable, and if it is, whether this flourishing should be equated with the satisfaction of the will and a celebration of its products, including the modes of thought that it inspires. After all, it is possible to think of ways of living where flourishing has consisted in subordinating our will to what is perceived to be some greater purpose or good – whether it be in a parent's attentiveness to the needs of their child, or a believers' obedience to their god.

Clearly, this opens the general – and very large – question of what is to count as true human flourishing: what sentiments and values should inform it? In the chapters that follow, I will offer a contribution to this debate. But in the present context, such considerations also raise the important issue of what a wilful attitude does to our relationship with the world, and particularly with nature. There are powerful arguments to support the claim that it is precisely the attitude of masterful manipulation towards the natural world that has led to our current environmental predicament. For example, in his influential *Steps to an Ecology of Mind* (2000 edition, p. 451), the anthropologist Gregory Bateson notes how when faced with a changing variable we tend to focus on modifying our environment rather than ourselves. With increasing technological power, yet continued inadequate grasp of the dynamic interplay of feedback systems in nature – in part a result of a lack of a felt sense of nature's integrity (of which, more anon), inevitably the outcome of this focus is frequently deleterious in the longer term.

Today, perhaps the most pervasive expression of an underlying attitude of mastery is exhibited in the strength and character of the consumerist economic motives that dominate Western society and that have led to widespread exploitation and despoliation of the natural world. The commodification of all (for example, as "natural capital", "human capital") is a clear expression of the motive of mastery – one that operates by habitually externalizing collateral effects that lie outside the chain of "most efficient" production. Any holistic understanding of the world is undermined and nature is set up as a pure resource for human consumption. The resulting lack of truly systemic understanding and the aggressive instrumentalism that holds sway in its stead could, as it gathers pace, hardly fail to cause harm both to the delicate natural equilibria in which human existence is embedded and to any fully humane frame of mind that could be properly sensible of this. It is in this sense that it is appropriate to speak of a prevailing metaphysics of mastery, for here we are installed in a particular (deficient) reality that remorselessly works to exclude anything that lies beyond its purview – especially any sources of intrinsic value that transcend the human will and could prove refractory to its demands.

To amplify the response to a previously raised question: the reason that quasi-mechanical portrayals of the natural world have the ascendency that they do is that they render the world as something that in principle can be controlled, mastered, whereas the recognition of spontaneity, fluidity, and otherness of purpose pervading the world disturbs this. So installed, it becomes increasingly difficult for us properly to address the issues to which the original advocates of sustainability wished to draw attention. Indeed, as the idea of sustainability becomes modulated through the metaphysics of mastery and articulated in terms of consumerist economic motives of increased efficiency and yield, it becomes part of the problem rather than the solution. In this guise, it both veils the real issues and legitimates attitudes and behaviour that continue to be ecologically destructive. When nature is perceived as purely a resource, any adverse consequences of exploiting it appear simply as needing to be fixed by either current or future technologies. Essentially, any problems are taken to lie not within the human will but within those aspects of nature that prove to be recalcitrant to it.

It is important to note here that mastery, in the sense in which I now intend it, should not be equated with instrumentalism in general; it refers to a specific kind of instrumentalism – one that sets us above, and therefore alienates us from, nature. To take a simple example, a hungry person seeing a cluster of berries as a source of food can be said to be taking an instrumental stance. The berries are viewed in terms of their ability to satisfy a current human need. But far from alienating us from nature, such an experience embodies a dependence on nature. The berries can be present to us as a gift (for which thanksgiving would be appropriate) and our relationship with them is an instantiation of an intimate connectedness with nature. It speaks of a belonging in nature.

The experience of metaphysical mastery is of a very different kind. This mastery (that is often tacit) is not primarily interested in simply seeking to utilize some aspect of the environment as a source of sustenance. Its primary objective is to make all subject to the human will. It sees everything as a potential resource regardless of whether there currently exists a relevant need to be satisfied (for this can always be manufactured) and seeks to convert everything, both in thought and in actuality, into something that can be possessed and ordered so as to be totally compliant with this goal. With the ascendance of this superordinate motive that involves a standing above and apart from things in order to assess their potential use, there comes a withdrawal from original experience that is coming to characterize the late-modern period. Expressions of this withdrawal include, for example, those now powerfully distilled in the narcotic hold in many lives of electronic connectivity, digitally constructed virtual realities, and, with technologies currently being developed, the likely burgeoning of digitally augmented realities. These all distance us from direct acquaintanceship with nature. In varying degree, they share the characteristics of Baudrillardian "hyper-realities" that have eddied off into a space of their own and that no longer have a proper external reference. Such hyper-realities have lost connection with any profound reality such as raw nature or a deep sense of the human condition and exist only as evermore pure free-floating simulations – much as when gossip takes off from reality and replaces it.

Preoccupation with such hyper-realities seems to intimate (and to fan) the previously noted deep and destructive disdain for reality as given. In her Prologue to *The Human Condition* (1998, pp. 1–3), Hannah Arendt noted how the launch of the first Sputnik was greeted in terms of its being "a first step towards escape from man's imprisonment to the earth". Here, what has been given is characterized only in terms of being an unwanted constraint – something that obstructs or thwarts human endeavour rather than what grounds and nourishes it, and sets proper boundaries to its aspirations. Under the metaphysics of mastery, the will can accept no boundaries that it has not itself imposed and can only be content with its own products that facilitate the exercise of its will and reflect back to it its ingenuity and power. The character and the implications of an accelerating preference for the artefactual in many areas of life are issues to which I will return. At present, it is necessary to consider some likely rejoinders to the critical stance being taken here towards the exercise of the will and ideas of boundless human aspiration. Does not this stance diminish the human spirit – and in specifically educational contexts might it not cabin students' expectations and visions of what might be and what they might become?

Taking the more general point first, there seem to be many occasions when the exercise of the will is a *sine qua non* of the indomitable human spirit and is to be wholly applauded. Consider the example of someone afflicted with a debilitating disease but who through sheer willpower overcomes the obstacles that they face, or an impoverished farmer who fights for her land and freshwater against large mining companies that are destroying and polluting the local environment. Here it is the exercise of the human will that carries each of them through adversity. It seems to me that considering examples of this kind helps to elucidate an important point: it is not the exercise of the will *per se* that is being questioned, only the exercise of the will that has become overweening because it looks only to itself, seeking to bring everything else under its sway and in so doing remains oblivious to the normativity, integrity, and intrinsic value that things in nature possess. This means that in the second example, it is the will of the mining company not the farmer that is the most obvious candidate for criticism, for, as described, the farmer seeks to preserve and protect while the mining company seeks only to exploit.

The point at stake is that our aspirations need to be tempered by this regard for the otherness (and well-being) of nature – indeed, it is partly this tempering that makes them genuinely high aspirations. There is nothing noble about the unfettered will. A will that is entirely enmeshed in itself, oblivious to the needs of all else – the expression of which is felt as just so much extraneous resistance to be overcome – is pathological. Nobility occurs only with a degree of subservience to something beyond the self-given that is recognized as having a greater call than the self. This has important implications for the educational case. The rhetoric of raising expectations and encouraging an attitude that everything is possible needs to be properly tempered in a culture where human self-gratification through material consumption has become a prime goal in life. It is not simply that the planet cannot support this on the scale that is now emerging, it is rather that if previous argument for normativity in nature is accepted, there is an ethical dimension that needs to be recognized. The question has to be raised as to what we are *entitled* to expect from nature, and how our expectations need to be moderated in the light of this. Part of this issue will be revisited when consideration is given to the idea of *nature's* needs in the coming chapter on ecological justice. For the moment, and having responded to some important criticisms of the notion, I would like to draw together some key strands of the idea of a transcendent nature and its way of "being" by considering the idea of nature's selfhood.

The "selving" of nature

Implicit throughout the account of nature that I have been developing is a sense that things in nature have their own integrity and that this needs to be respected. They are not simply there *for us*. They have their own existence, in some cases lives, and in some ways can be regarded as there for each other. This dimension of nature's existence has been expressed by the great nature poets, and in particularly explicit form in some of the poems of Gerard Manley Hopkins, to whom I have previously made reference. In the following lines, he expressed his strong sense of things in themselves and *for* themselves.

> As Kingfishers catch fire, dragonflies draw flame;
> As tumbled over rim in roundy wells
> Stones ring; like each tucked string tells, each hung bell's
> Bow swung finds tongue to fling out broad its name;
> Each mortal thing does one thing and the same:

Deals out that being indoors each one dwells;
Selves – goes itself; myself it speaks and spells,
Crying What I do is for me: for that I came.

Here the interiority of things in nature is proclaimed – what Hopkins has called their "inscape", the thing itself in its own special nature. Matthew Farrelly (2019) suggests that what we might call some thing's inscape is "its inner landscape of being, its soul, or its essence". Now consideration of this notion of the inner being or selfhood of things in nature raises some important considerations that can help to clarify the account of the occurring of things given in the previous chapter. As developed there, great emphasis was placed on the idea of the thing's being consisting in its native "occurring". On this basis, I suppose that it would be possible to interpret the being of, say, the beech tree as purely phenomenal in the sense of being constituted exclusively in a sensual occurring in interplay with that of its neighbours. There is a sense in which this is true, but what needs to be recognized is that the interplays described are not sensory in the purely physical sense.

To begin with, on the account that I develop the "thisness" or *haecceity* of the tree consists both in its individual here and now appearing from moment to moment and its being a particular kind of tree. In classical philosophy, this latter aspect of the existence of a thing was conceived as its "essence" or "quiddity". Hence, the quiddity of the tree was taken to consist in those essential characteristics that make it a *beech* tree as against some other kind of tree, such as an oak. And today, for reasons elaborated previously, these characteristics are likely to be set out in terms that reflect scientific proclivities that attempt to define things in accordance with pre-specified "rational" categories. This would ossify the tree in ways previously described. But this need not – and on the account supported here – certainly should not be the way in which we seek to understand a natural thing's essence. In their unique occurring as beech trees, they have myriad fluid and non-pre-specifiable possibilities – a nascent potentiality that is theirs as beech trees. The key point here is that in its "thisness", the tree does not appear simply as an example of its kind, but as this unique beech tree embracing and embraced by a unique and ever emergent place. In this sense, "thisness" can be taken to refer to this particular beech tree in its here and now occurring – what one might refer to as its "suchness". Here perception is all; there is nothing beyond appearing. But this is not to deny the significance of its essence, rather it deepens it by seeing it as interfused with its occurring being.[1] This holds because, as previously noted, on the view being developed here, perception is not purely sensory in the sense of apprehending purely physical objects, but necessarily includes sensing or intuiting significances that lie beyond this dimension – yet that tacitly inform it through and through.

Instantiations of what one might term the "essencing" of essences in nature can be foregrounded in works of art. For example, think of some of Vincent Van

Gogh's paintings such as *The Large Plane Trees at St Rémy* (1889) or *Tree Trunks in the Grass* (1890). In the former, amongst other things, the sheer massiveness of the limbs of these particular trees is brought to our attention as an inherent part of their being that in part makes them the trees that they are, and in this sense defines them. But it in no way reifies them, for this massiveness is itself unfathomable and always becoming. Similarly, the gleaming many hued tree trunk presented in *Tree Trunks in the Grass* brings out powerfully the character of its bark – indeed, one might say manifests the "barkishness" of its bark – that again expresses something of both the essence of this tree and of the essence of bark. And it does this in a way that while bringing to light also preserves the mystery of this aspect of nature such that it cannot be fully articulated and captured in a definition, and thereby ossified. The power of art to renew and extend our perception of the natural world is a topic to which I will return in the final chapter on ecologizing education. For the present, I return to the matter raised earlier of a sensing that is not purely sensory in the traditional sense.

The following observation made by Maurice Merleau-Ponty (1962) on the nature of colour perception is apposite:

> We shall not succeed in understanding perception unless we take into account a colour function which may remain even when the qualitative appearance has been modified. I say that my fountain pen is black, and I see it as black under the sun's rays. But this blackness is less the sensible quality of blackness than a sombre power which radiates from the object, even when it is overlaid with reflected light, and it is visible only in the sense in which moral blackness is visible. The real colour persists beneath appearances as the background persists beneath the figure, that is, not as a seen or thought of quality, but through a non-sensory presence.
>
> *(p. 305)*

If, here, we interpret his reference to a "non-sensory presence" as denoting a presence that is sensed but not strictly physically present, we can see that in essence, the thing should be regarded as an inter-sensory entity. This includes, now, the sensing of non-physical properties such as ambiance and kind that, too, arise in the occurring of the beech tree. Here kind is experienced not now as definition, but as a call of the sort expressed in the lines of Hopkins' poem cited previously. In receiving it, we are intimately involved as embodied individuals. So experienced, the thing is what is taken up by our gaze or, indeed, by our movements, and these latter are so intimately interfused with what we perceive that they can be experienced as "questions" to which these things provide a fully appropriate "reply" (p. 317). Merleau-Ponty goes on to reinforce this quality of mutuality in perception by saying: "what I call experience of the thing or reality . . . is my full co-existence with the phenomenon, at the moment when it is in every way at its maximum articulation" (p. 318). With this reassertion of the subjective in all perception, it follows that essentially perception is not a science and that to allow science to colonize our understanding of our most elementary relationship with the world and what binds us to it is to distort our apprehension of the "antepredicative being to which our whole existence is polarized" (pp. 321–2).

With considerations of this kind in mind, it can be seen that a proper account of "thisness" requires recognition of two dimensions: the non-sensory presence of the "treeness" of the tree, and the individuality of the native occurring of this particular tree in this particular moment. Perhaps something of this distinction – this bi-fold "thisness" – can be caught through drawing an analogy with persons as when we relate to them as a "what?" and a "who?". The former might elicit characteristics of what it is to be human, and the latter the individual being of a particular human being as they stand before us – as sometimes summonsed up by calling their name. Each is necessary for the intelligibility of the other: "whatness" arises in the context of an individual's standing forth, "occurring", and also it is implicit in this event. But this analogy only takes us so far, for it has an interrogative tone to it that can be absent from the kind of receptivity that I am suggesting. To a degree, this would close down the openness of this receptivity, delineating the manner in which the thing can stand forth. But it makes the point that the "what" and the "who" – the essence and the particular in its here and now of appearing – are intimately intertwined. Perhaps this view can be summed up by harking back to the analogy with human beings: individual things in nature – such as the beech tree – have something akin to a personality. It has individual characteristics that are a blend of those of its kind and those of an individual uniquely situated self. Experienced in this way, and through a developed and attentive acquaintanceship with the emplaced tree, we can become familiar with its individual traits and nuances – as contrasted with some set of objective properties.

The "really real"

Having amplified these aspects of the standing forth – occurring – of things themselves, I now turn to some broader summary considerations concerning the reality of nature and what I mean by calling it "transcendent". These are preparatory to exploring the idea of environmental consciousness that is the subject of the next chapter.

Foregoing argument has sought to establish the necessity of non-sensory perception – what we might now call "intuition" – in our experience of things in nature. This provides the ontological space for the experiences that we can have of things in nature as having their own integrity and fulfilment that is deserving of respect. As previously discussed, clearing this space has long been a powerful motive in the work of the Romantics. This, in turn, can be seen to reflect a more general intuition that there is a sense in which nature has its own mysterious purposes and that, although ultimately mysterious, there is a level at which we have a natural sympathy with them. Indeed, when not preoccupied with other matters, there are occasions when we can sense ourselves as a part of this telos and recognize its authority: it is part of *our* nature.

What is being affirmed here – and what it is essential to affirm – is a respectful intimacy in knowing nature in which it is brought close, but in which at the same time its autonomy, otherness, is preserved. This makes it clear that nature conceived as some scientific system is at best highly partial and in many ways it misses the point altogether. Nature is not essentially an energy system, nor a deterministic causal network, nor an instantiation of abstract laws. All such intellectualistic description represents a systematization that subverts both our original experience of it and the character of its transcendence. Notwithstanding the senses in which *concepts* of nature (scientific or otherwise) are social products that may vary over time, there is a certain constancy in our elemental *experience* of nature. Nature befalls us; it is forever nascent, inherently largely undisclosed and in living interplay. Involvement in the vital occurring of nature is imbued with a sense of the withdrawn – of that from which things arise, of aspects out of view, of what was and anticipation of what is not yet. And the draw of this withdrawn can sometimes be more sharply felt than that which seemingly is immediately present before us. It makes a call upon our thinking, leading it on, constantly alerting it to the possibilities of an exploration of the unknown. Attendance to this is thinking in a demanding sense. Indeed, it could be considered to provide a paradigm for the kind of thinking that education should cherish – for as we lose our sense of the withdrawn, so we lose touch with a latent reality that can refresh and inspire.

Finally, in drawing this chapter to a close, it will be helpful to clarify philosophically one further aspect of the sense in which I am now claiming that nature is transcendent and what this implies for articulating our experiences of it. There are two dimensions to this. The first can be expressed in the following way. In speaking of natural phenomena in their own being, I have referred to them as "things themselves" rather than "things *in* themselves". This locution is intended to distinguish the kind of transcendence that I have in mind from a form of transcendence that has a long history in philosophy and that found particularly powerful expression in more modern times in the thinking of Immanuel Kant.

Classical philosophers such as Parmenides and Plato drew a distinction between appearances: "phenomena" – that are accessed through the senses, and some underlying or higher reality – "noumena" – that can be accessed only through intellectual intuition supplied by a refined form of reason. The world of phenomena is one of constant transience and change; nothing endures, and to a degree it can be subject to the whims of humanity. By contrast, noumena are eternal and unchanging, far removed from the everyday world of corruption and decay, and knowledge of them – perhaps, even some sort of submissive participation in them, in the sense of reflecting what they intimate – represents the ultimate fulfilment of the intellect. This instantiates knowledge of a supreme reality that is always imperfectly reflected in the appearances that constitute the phenomenal world. We might say that for classical philosophers of this persuasion, it was knowledge of the noumenal that constituted knowledge of the "really real", and clearly, it would be knowledge of this kind – of, what is most fundamental, what is real, true, in the fullest sense – that should form the basis of our lives and decision-making on important matters.

There is a respect in which, echoing the views of such notable classical philosophers, Kant also allowed the possibility that there exists a reality that underlies all mere appearances: a world of noumena that he referred to as "things *in* themselves". Members of this world are transcendent in the sense that they, as it were, give off appearances, but in principle lie beyond what can be perceived. However, contrary to Plato's view, Kant in *The Critique of Pure Reason* argued that we can speak of the noumena only in a *negative* sense of referring to things that are not objects of a sensible intuition. This negative emphasis arose because Kant claimed that we neither possess a non-sensible (i.e. purely intellectual) intuition, nor can we understand what such an intuition would be like. Hence, the noumenal are construed as constituting an ultimate reality that is mysterious in the sense that we can know nothing whatever about it. This form of transcendence is so complete that the things taken to enjoy it are completely isolated from human experience and life. Their existence is a matter of sheer inference and their true character is a matter of sheer ignorance.

In stark contrast to this, the transcendence that I intend by the designation "things themselves", while certainly referring to an aspect of things that lies beyond human construction and authorship (they are "themselves"), does not refer to things that are so "*in* themselves" as to be beyond all knowledge and sensory perception (as with Kant), or indeed, have an existence that is completely independent of human participation (as with Plato). As previous discussion makes clear, "things themselves" refers to things that are always showing themselves, but never *comprehensively*. They have aspects that are always withdrawn, their arising is always out of what they are not and what is not currently present. And while in this sense they have a life of their own and countenances that we will never witness – and therefore in their entirety they can never be adequately summed up by us – nonetheless they occur for us. And, as has been previously illustrated, there is an important sense in which we participate in their occurring. The central point here is that the transcendence of which I speak says nothing about some other non-phenomenal, non-experiential world; it is entirely located in phenomenality – but now, compared with classical conceptions, an enriched notion of phenomenality – and our aspiration should be to attend to it in all its facets, and to detect themes that run through it. This view locates *being* in this phenomenality and holds it to be primordial in the sense that it is here that things in nature self-arise and that careful attentiveness to what occurs – that is attentiveness that submits to what is before it without ulterior motive – reveals things simply as they are.[2]

Here there is an important sense in which the "really real" communicates something of itself to us and that is articulated better through poetry than science. In this regard, Michael Chang (2020) has suggested that the terse Haiku poetry of Basho that attempts simply to say what is, without flowery or sentimental embellishment (that, perhaps, is sometimes a feature of Romantic poetry), can be a true expression of nature's voice. He gives the following example:

> Ah – speechless before
> These budding green leaves
> In blazing sunlight.

Clearly, this is much leaner than, say, the poetry of Wordsworth or Hopkins. Chang cites Basho as characterizing the poem in its brevity as "spare, clean, swift as a boning knife" and speaks of it as conveying a perceptual acuity typified by encounters with suchness. Regarded exclusively in terms of voicing nature (rather than any pretence of evaluating the works as poems), should we say that because it is a simple expression of the sensory phenomenality of nature that it is truer to nature? This returns us to a key point discussed previously, and one that is very relevant to the forthcoming discussion of nature's voice to be undertaken in Chapter 6. The fundamental question to which we are returned here is whether nature's phenomenality is exclusively sensory. I argued that when Hopkins spoke of the "selving" of nature, this involved something beyond pure sensory perception as conventionally understood. In addition, it repudiates the scientistic claim that any non-sensory element necessarily would be the product of a human projection rather than anything emanating from nature itself. While the particular Haiku cited does, indeed, express nature – and in a very significant respect – there are facets of nature as the self-arising that are additional to what it conveys. When Hopkins speaks of each mortal thing crying "What I do is for me: for that I came", or when, as I have claimed, we detect normative reverberations emanating from nature, clearly (from a classical point of view) a non-sensory form of intuition must be in play (although, as I have argued, it should not be regarded as independent of simultaneous sensory intuition). Yet it is a constitutive part of our experience of things in nature, of their occurring for us. It is part of our sensory experience in that we *sense* it, just not through the five senses as traditionally conceived.

This reference to poetry, that of course is a human artefact, brings me to the second element of clarification of what I mean by nature's being transcendent. This can be expressed by referring to poststructuralist views that reject the idea of a meaningful reality that lies beyond "the text" and to which language refers us. This would seem to cast nature as essentially a social construction, possessing no intrinsic autonomy; it is simply a product of how it is characterized in the text. Nature is, in its entirety, what the word "nature" (and its cognates) mean in the context of some ongoing text. Claims of this sort raise very large philosophical issues, but for present purposes perhaps they can be circumvented by accentuating the point that the intention of the current argument is not to identify or posit some pristine underlying reality that exists in some absolute sense and that, for example, is antecedent to and presupposed by language (or "the text"). Rather, the aim is simply to draw attention to what is given in our experience of things (in the sense of what our full attentiveness can reveal, as elucidated previously) regardless of what might be considered to be its ultimate genesis in scientific, social, or even philosophical terms. And also to explicate the claim that what is given here nonetheless is original in the sense of being alert to, and hence grounding us in, the play of being of things themselves. It follows that views of the kind previously described in this chapter, and that I have dubbed "enemies" of transcendent nature, should be understood neither as being capable of undermining this original property, nor the authority that it is capable of emitting.

Notes

1 This continues to hold true when we do not know that, say, the tree before us is a *beech* tree. Indeed, even when the thing before us is completely strange, it remains experienced as a thing of some *kind*, and whose qualities can emerge as such.

2 It follows that the "phenomenality" of which I speak is not to be equated with traditional phenomenalism of the kind defended by, for example, A. J. Ayer, and that focuses on objects in their pure "physicality". Here it is held that physical objects can be reduced to sensory experience – that is "sense data" – without remainder (or, putting it linguistically, that physical object statements can be analysed exclusively in terms of statements describing sensory experience). Rather, it admits aspects beyond their pure physicality to the being of things, to phenomena, and in this sense collapses the traditional dualism between what appears (to the senses) and what is thought, but not now where thought is construed in terms of abstract reason, but rather what is concretely sensed and *felt*.

4

ENVIRONMENTAL CONSCIOUSNESS

Intentionality and ecstasy at the centre of human being

Preceding chapters have sought to develop an understanding of the being of nature in its native occurring and to identify some of its essential qualities. It is now time to say something more about the consciousness that experiences nature so described and to explore the character of the relationship between this consciousness and the nature that it experiences. Understanding this relationship is the key to developing an idea of what an appropriate response to the current environmental crisis would require and lies at the kernel of the idea of ecologizing education. The latter follows from the fact that ultimately it is with the quality of consciousness that education is chiefly concerned and upon which it hopes to make an impression, bring about some change for the better. Elucidating this consciousness requires an exploration of the meaning of selfhood and the character of the idea of there being a human essence because, as the preceding chapter sought to demonstrate, the occurring of things in nature happens primarily in the space of individual embodied human subjectivities – that is, selves.

Towards ecological selfhood

In his extensive work *Ideas of the Self* (2005), Jerrold Seigel argues that the basis of selfhood in Western culture has been sought primarily along or within three dimensions – these he terms the bodily or material, the relational, and the reflective. On this schema, the bodily dimension refers to the corporeal existence of individuals that makes us palpable creatures shaped by the body's needs and proclivities. The relational dimension arises from the social and cultural interactions that shape us. And the dimension of reflectivity derives from the human capacity to make both the world and our own existence, including our consciousness, objects of our active regard. It can be seen that the account of our awareness of the occurring of things in nature given in the previous chapter instantiates these dimensions in some respects, but not all. For example, there the relational dimension was articulated rather differently, the emphasis being put on our participation in the natural rather than the social, and at times this participation was characterized in a way that tended to de-emphasize the subject's self-awareness (reflectivity) in favour of being at one with the phenomena with which it is engaged. But even here some level of self-awareness would be hovering at the margins in that the subject would be able to recall that it was *they* who had the experience and bear some responsibility for it. In this and succeeding chapters, this orientation towards Seigel's three dimensions that results from a focus on the natural will receive further amplification, but for the present – and by way of preliminary to this further amplification – it will be helpful to consider some social aspects that frequently have been taken to determine consciousness and hence condition its openness, including its openness to nature.

In the social/political context, there has been a long history of conceiving the self, not as some separate autonomous entity – as has been attributed to some forms of liberalism – but as essentially relational. This has sometimes resulted in subjectivity becoming understood largely as merely the creation and ongoing reflection of external influences. A classic example of this is Marx's depiction of social relations – particularly those emanating from class division – as determining consciousness and, indeed, perceived bodily needs. In the educational context, Basil Bernstein's (1965) still influential analysis of "restricted" working-class and "elaborated" middle-class language codes that are taken to fundamentally orientate and structure the subject's thinking can be seen as an elaboration of this. In more recent times, the general view is illustrated in varying degrees through, say, reference to ideas of "subjectivation" through the performative activities of others (see, for example, Butler, 1997) or the subject's coming into presence only through the recognition or the call of others (see, for example, Biesta, 2006). With the former, the basic idea is that we become self-aware through perceiving how others respond to us. The ways in which they behave towards us and the public ways in which they speak about us are "performative" in the sense that they do not merely describe us, but perform the function of constituting our sense of who, as an individual, we believe ourselves to be – our sense of selfhood. It is held that in reality, we have no pre-existing inner individual essence that expresses itself in the world, rather our subjectivity becomes a product of what others say and do in relation to us: we are "subjectivated" – *given* our subjectivity from outside. With the latter, at one level we have something of a truism: our subjectivity only becomes part of the public world as it is recognized by others. But there is a stronger version that holds that our subjectivity is called forth by the moral demands of others that challenge us to respond to their needs. As will be seen, suitably modified, this approach to selfhood has the potential to resonate strongly with what I argue to be the proper character of our relationship with nature.

Views of this kind often derive from the philosophy of Emmanuel Levinas and his idea of the "face" as the living presence of another person: the way in which the other presents him/herself that infinitely exceeds any idea or image of them that we might have, and that invites us into relation. Levinas argued that our subjectivity emerges only through an encounter with the other and in this sense subjectivity is heteronomous rather than autonomous as was held by Enlightenment views. Hence, he holds that subjectivity exists in a condition of existential indebtedness to the other and bears a responsibility to it – a relationship that he has famously described as "older than the ego, prior to principles" (Levinas, 1981, p. 117) because in his view it is a condition of the ego's existence. As Ann Chinnery (2018) puts it in her exegesis of Levinas' views:

> [F]or Levinas, one comes into being as a subject already in relation with another whose irreducible alterity (exemplified by the face) means that she or he cannot be assimilated or reduced to some version of oneself. So, rather than being connected to the other by way of essential similarity or a common humanity, we are called to a different kind of kinship. . . . [O]ne is called to be-for-the-other in response to the face of the other in all its vulnerability, suffering, and destitution. I am called into being as "I" not by the force of a powerful counter-ego, but by the somewhat paradoxical "force" of vulnerability and fragility.

The authority of vulnerability is a key idea here, as is a repudiation of ideas of the self, frequently attributed to Descartes, in which it is characterized as existentially self-sufficient and as lying at the kernel of meaning-giving, positing an external world whose reality it then needs to prove. Here the flow of meaning is viewed as being from the self to the other; the external world is endowed with significances by the thought and activity of a meaning-giving subject.

Now undoubtedly there is a good deal of truth in ideas that seek to qualify this Cartesian view – indeed, to overturn it in some crucial respects. For example, clearly our sense of self will be influenced by the ways that others treat us and behave towards us, and new aspects of ourselves can be revealed – be brought into being – through our intercourse with others. Furthermore – and very pertinent to the views being developed in this book, there are some clear resonances with the emphasis that is laid upon the presentation of otherness in its infinite alterity in Levinas' account and the otherness and mystery of the native occurring of things in nature that feature in my account of the phenomenology of nature given in previous chapters. More will be said on this in chapters to come, and also on the attempt, exemplified by Levinas, to invert the modernist conception of subjectivity as possessing a kind of independence and sovereignty – for clearly, to the extent that the self is conceived as being called forth in its relations with the world (the "other") rather than being conceived as some complete pre-existing autonomous entity that has dealings with it, this will have extensive implications for any educational project, including that of ecologizing education.

For the moment – and returning to the present exploration of the nature of selfhood – it needs to be noted that it is one thing to say that some aspects of one's subjectivity can be called forth through, or indeed initiated by, a relationship with the other, and quite another to say that one's subjectivity is entirely, or primordially, constituted by this relationship. Furthermore, it is not at all clear as to why one should feel, universally as it were, a sense of infinite responsibility towards the other as against sometimes feeling sympathy or empathy. A feeling of universal responsibility towards individuals that we encounter hardly seems to be a feature of human experience even when it is at its most receptive, and to say that this may be so, but that one *ought* to feel such responsibility would contradict Levinas' claim that it is primordially constitutive of the self, for it would no longer be "older than the ego". Indeed, now considering the situation as a whole, a too exclusive a focus on the role that relational features of this kind have in the constitution of the self can lead to the danger of any idea of an individual's authentic human essence becoming dissipated across an interminable range of external agencies (Bonnett, 2009a). Here, if there is only what others make of you or evince in you, then individual subjectivity loses the interiority (and perhaps persistence through time) that makes it a genuine subjectivity in the first place. For example, in this view, it would seem that essentially we simply read ourselves off from what others say of us and how they treat us. Here, our individual selfhood is in severe danger of becoming the product of what Martin Heidegger (1962) has termed "hearsay": the beliefs, thoughts, and feelings currently circulating in what people say and that are not necessarily grounded in anything more than gossip. On this account of its constitution, the self has no intrinsic integrity and ideas of, for example, organic personal growth – and, indeed, personal responsibility – that have long been central to ideas of human subjectivity or selfhood become highly problematic.

Yet, on the other hand, to take the view that there is such a thing as an authentic human essence and that the kernel of a person's subjectivity – its integrity – lies in this and therefore beyond, or even is antecedent to, the recognition or performative power of others would seem to be in danger of falling into a form of solipsism and to involve an attempt to provide an abstract definition of what it is to be human that would reify it. The problem here is that in attempting to lay down specific criteria for what is to count as human, the idea of an authentic human essence can be portrayed as placing arbitrary restrictions upon human potentiality that in turn involves a colonization of humanity by the views of the intellectual/religious/political elites who decide the criteria. A gross example of this would be the Enlightenment elevation of European bourgeois reason as central to what it is to be properly human and thereby universal arbiter of good thinking. This has been aptly dubbed by Robert Solomon (1980) a "transcendental pretence".

However, there are alternative ways of acknowledging the relational dimension of human being, and that emphasize the involvement of subjectivity with an environment, yet that do not jettison the idea of a self that possesses some sort of internal integrity and whose characterizing aspects can be articulated to some degree. It seems to me that one such notion of authentic human essence flows from a formulation of the idea of consciousness derived from medieval schoolmen and reintroduced into modern philosophy by Franz Brentano in the latter half of the nineteenth century: the "intentionality thesis". I will develop the view that here, with some modifications, can be discerned a view of human being that holds the possibility of developing a notion of human essence that reveals ways in which human consciousness is ineluctably environmental, and thereby exists in an internal relationship with nature. It will be argued that this presents us with a primordial idea of sustainability that in turn has the potential to reorientate our understanding of education and what might be involved in ecologizing it.

Consciousness as intentionality

In his *Psychology from an Empirical Standpoint* (1995 edition), Franz Brentano employed the idea of intentionality as a way of distinguishing consciousness from the merely physical. He interpreted intentionality as "relationship to a content, the tendency towards an object" that is immanent, i.e. contained within consciousness. This idea of consciousness as essentially directed upon an object, that is, being "minded", subsequently was taken up by Edmund Husserl (2001 edition) and seminally modified by his argument that the things to which consciousness is directed – what he termed its "intentional objects" – are not contained within itself, but are *transcendent*. For example, when we desire something such as a new coat, we do not desire something that is already within consciousness, say as an image or an idea, but an actual coat (perhaps, one currently displayed in some shop window) whose existence lies beyond any individual consciousness. Furthermore, as Martin Heidegger (1962) makes clear in his rejection of Husserl's developing transcendental idealism, we experience these transcendent objects as always already existing in a public world that they share with us. For example, if they are terrestrial objects, in their physicality they are subject to the same physical phenomena as we are in our physicality, such as changes in temperature and humidity. It is the same stiff breeze and slanting rain that causes the leaves of the nearby tree to rustle and glisten as causes us to pull up the collar of our coat. These transcendent objects constitute the things that we negotiate, handle, use, contemplate, and so forth: they populate and structure our world in a web of significances to which we and they are bound, and in which we both exist. Such "worldliness" – participation in this world of shared significances – is fundamental to their intelligibility and our understanding. It follows from this that human consciousness is ecstatic in the sense of existing in a constant (and complex) motion of standing out towards things beyond itself in an intelligible world – a world that in principle is publicly shareable.

This relating to things occurs through both thought and action; indeed, in our ecstatic participation in the world, the two are often indissoluble. In this sense, consciousness, also, is ineluctably worldly – and hence, we can say, *environmental*. Its existence consists in its relating to an environment – indeed, *the* environment: the environment that in some sense we all share and in which our lives are embedded. And the internal integrity of an individual consciousness can be conceived as a constantly evolving genealogy of intentionality: of relating to things (both through thought and action) whose existence is embedded in this shared world. Here past acts of intentionality condition, but far from determine, future acts (Bonnett, 1978).[1] This is to say that an individual's "intentional history" plays a part in both situating them and giving them their own outlook – the unique locus from which they understand and engage with the world. This intentional history is the food of memory and is sustained by it, and possesses the quality of "mineness" that constitutes our sense of selfhood. But clearly, although in a sense pre-existing, this self is by no means fixed. Rather, it exists as so much potential that is capable of being realized in myriad and often unpredictable ways, depending on what it encounters. It is in constant process of being modified and of evolution through its encounters.

Relevant to this understanding of the intentionality of consciousness that situates it in a shared world, Hannah Arendt opens her influential book *The Life of the Mind* (1978) by exploring the claim that what all things in the world have in common is that they *appear* and hence are *meant* to be perceived. As she puts it:

> *Being and Appearing coincide.* Dead matter, natural and artificial, changing and unchanging, depends in its being, that is in its appearingness, on the presence of living creatures. Nothing and nobody exists in this world whose very being does not presuppose a *spectator.*
>
> *(p. 19)*

Referring this back to Husserl's development of the intentionality thesis of consciousness that we have previously discussed, she claims that:

> Husserl's basic and greatest discovery . . . is the fact that no subjective act is ever without an object. . . . [O]bjectivity is built into the very subjectivity of consciousness by virtue of intentionality. Conversely and with the same justness, one may speak of the intentionality of appearances and their built-in subjectivity. All objects because they appear indicate a subject, and, just as every subjective act has its intentional object, so every appearing object has its intentional subject.
>
> (ibid. *p. 46)*

This is a seminal thought. Taken as an amplification of the fundamental intentional nature of human consciousness and the character of its objects, it leads to a further disclosure: there is an important sense in which consciousness is ineluctably involved in sustainability. In requiring a subjectivity (consciousness) for their *appearing*, that on this account coincides with their *being*, objects not only indicate a subject, but subsist through an ontological relationship with it. Hence, consciousness is confirmed as the place where things appear. This means its essence is to let things *be*. In doing this, it *sustains* them as the things that they are. Equally, and very importantly, it is sustained *by* them. The life of consciousness is sustained by the things that appear to it: its being consists in its participation in a relationship to the being of the things that appear to it. Things exist in their *meaning* to be perceived, and consciousness exists in its *meaning* to perceive them. Hence, consciousness and its objects are always already interfused: there is an *internal* relationship between them in which the agency of both is operative. In effect, this latter is precisely the message that the concrete examples of the woodland dell and the upland stream explored in Chapter 2 bespeak and illustrate. Our own being is enlivened by our ecstatic openness to both the otherness of these natural places and our sense of what is a fitting response to what we encounter there as a result.

If it is granted that such ecstasis (intentionality) is, indeed, a definitive characteristic of human consciousness in the ways elaborated previously, an avenue is opened to examine further the significance of the contribution of the natural world to the life of consciousness. There have been many who have expressed the view that direct contact with nature makes an especial contribution to human well-being. As elucidated in Chapter 2, certainly this was the view of Romantic poets such as Wordsworth. More recently this position has been taken and developed by Richard Louv (2010) with his notion of "nature deficit disorder" and is also endorsed at the everyday level by the many who seek natural settings for their leisure and recreational activities. It is possible to give these expressions of a need for direct experience of natural places a psychological interpretation: it is a purely contingent empirical fact that some people feel this. What now arises as a possibility as a result of previous analysis is that this need is not only empirical but also *ontological*. I will argue that it is through an engagement with nature that the ecstatic character of human consciousness is most fully realized, and that therefore in this sense the phenomenological reality of nature is a key reality for our authentic being. Clearly, to the extent that this argument holds, it will have considerable implications both for the nature of education and for how we should understand the philosophy of education. There would be an important sense in which all education would need to be considered as environmental not only in a general sense that would include the social and the artefactual, but also in the specific sense of being nature-rich, ecological – attuned to the environment of the self-arising, the native occurring of things in nature.

For the purpose of elucidating how this internal relationship between consciousness and its environment connects the idea of an authentic human essence with participation in the native occurring of things in nature, I will develop in more detail the two themes identified previously:

1 The primordial character of human ecstasis as a kind of sustaining;
2 The significance of the experience of nature for human flourishing.

Human being as sustainability

In other work (Bonnett, 2004), I have proposed that there is a pregnant sense in which sustainability lies at the heart of human consciousness, and have argued that understanding this relationship is intimately entwined with understanding a foundational notion of truth. The argument can now be developed as follows. If it is proper to characterize human consciousness as intentional in the sense outlined in the previous section, it follows that the greater the range and integrity of the intentional objects in which it participates, the greater the richness of its own life. And because these objects are transcendent and therefore not to be conceived as exclusively the product of its own projections, primordially its stance will need to be one of receptiveness to what engages it. Here truth is conceived not in terms of the correctness or otherwise of statements, but rather, following Heidegger (1975), as an original revealing or disclosure of things that is antecedent to the making of statements and, indeed, that makes them possible. The judgements that statements involve require some sort of prior apprehension of the things judged.

There would seem to be echoes here of Aristotle's ideas of *nous pathetikos* and *nous poietikos*. Discussing these in the context of environmental issues, Pall Skulason (2015) elucidates *nous pathetikos* as a basic and enormous receptive power that we have that "makes us connect with nature and all its phenomena by taking on, so to speak, their forms within our own minds". *Nous poietikos* he characterizes as "the potentiality for producing concepts and objects which then may direct our further decisions and actions". It makes it possible to discern the various forms that we have already received from nature and to play with them within our minds so as to form all sorts of images and theoretical constructions. This activity of *nous poietikos* is based upon what *nous pathetikos* has provided. Skulason suggests that during the modern period,

> the intimate connection between *nous poietikos* and *nous pathetikos* has broken down, leaving *nous poietikos* to operate alone, ignoring the meaningful forms that come to us through *nous pathetikos* and ignoring the vital fact that we are ourselves the part of nature that makes nature aware of itself.

Here, then, the primary significance of receptivity is foregrounded as foundational to any further intentional activity of consciousness, and as grounding it in a reality that both enriches and authenticates its thoughts.

Interestingly, reasoning of this general kind can be found in philosophical traditions other than the phenomenological one that largely informs the current work. For example, although he makes no explicit reference to any intentionality thesis and the related arguments developed previously, something very like them can be interpreted as being present in Bertrand Russell's introductory text *The Problems of Philosophy* (1959). In the concluding chapter, "The Value of Philosophy", he suggests that true knowledge is a union of the Self with the not-Self. It is important to note that in the case of philosophy, this union is not to be understood as generating a body of definitely ascertainable knowledge as, say, with Plato. Rather, Russell holds that: "The value of philosophy is, in fact, to be sought largely in its very uncertainty" and the consequent speculation that "suggests many possibilities which enlarge our thoughts and free them from the tyranny of custom" (p. 91). In this way even the smallest and most familiar things in life can become strange and enlarged in their significance. This leads Russell to make a seminal point that reveals a key implication of the intentionality thesis when considered from an educational perspective:

> Apart from its utility in showing unsuspected possibilities, philosophy has a value – perhaps its chief value – through the greatness of the objects that it contemplates, and the freedom from narrow and personal aims resulting from this contemplation.
>
> *(p. 91)*

Clearly, there can be a debate here over what will count as "greatness", but the important point that he goes on to make clear is that such enlargement of Self does not occur through study that

> wishes in advance that its objects should have this or that character, but adapts the Self to the characters which it finds in its objects. . . . In contemplation . . . we start from the not-Self, and through its greatness the boundaries of Self are enlarged; through the infinity of the universe the mind which contemplates it achieves some share in infinity.
>
> *(p. 92)*[2]

Here we are invited to participate in a radical de-centring of self in perception, and in this fundamental sense of consciousness being attentive to what its intentional objects present rather than to pre-existing preoccupations, it is involved in a sustaining of things – a letting them be as the things that they are. For an individual, this is the basis of world formation – of truly being *there* in the world. And here, too, resides an original sense of truth: an apprehension of things as they are in their own being – that is, how they present themselves to an open and attentive mind. This resonates strongly with both the idea of truth as disclosure alluded to previously and also the idea of nature as inherently mysterious and requiring receptivity for its appearing. But it should be noted that this, in itself – and consistent with the view of primordial reality developed in the previous chapter – is not to posit some objective reality entirely independent of consciousness *per se*; rather it is to speak of a reality, in participatory relationship to which, primordially consciousness consists or lives. And, by the same token, it is a reality that occurs through consciousness, as the place where it can show up. Here, as it were, subject and object appear as poles of the relationship that is consciousness. The poles can be distinguished, but arise only as aspects of this original intentional relationship that constitutes consciousness, and that at base is one of mutual anticipation of the kind observed by Arendt.

In the previous chapter, I developed the idea that, contra scientific naturalism, if we take a phenomenological approach to the natural world, it reveals itself as rife with anticipation. As I will argue that anticipation constitutes an important element of what it is for us to be *in* nature, the senses in which we anticipate nature and that arguably *it anticipates us* need to be amplified. In order to provide a proper context for this, it will be helpful to return to the idea of the significance of place and the character of our emplacement – the ways by which we inhere in a place and are therefore receptive to it and its inhabitants. For it is in this context that the extent and character of the phenomenon of anticipation comes to the fore.

Emplaced consciousness

The account of the occurring of things in nature developed in previous chapters foregrounded a number of overlapping features that I took to contribute to the individuality of a particular thing in nature: ideas of neighbourhood; inherence; relationality; dynamic, reciprocal participation; mutual sustaining. It also emphasized that nothing that we encounter is unplaced. This brings to mind Martin Heidegger's remark that place is the "locale of the truth of Being" – by which I take him to mean that Being is unconcealed – i.e. things come to presence – not in some abstract uniform mathematical space-time framework, but in particular locally structured places that have features and ambiences that are uniquely their own. One never encounters such things as, say, a blossoming tree in uniform space, only neutered (perhaps scientifically or mathematically defined) objects devoid of phenomenological vigour and intrinsic significance.[3]

Consistent with previous discussion of intentionality, for a conscious being place-making – the reciprocal participating in the constitution of a neighbourhood and being conditioned by that neighbourhood – occurs in the space of *intelligibles*. This is to say that it occurs in a space composed of things possessing interwoven significances that can be – and to some degree always are – *understood*. As Michael Oakeshott (1972) once put it:

> Human beings are what they understand themselves to be; they are composed entirely of beliefs about themselves and about the world they inhabit. They inhabit a world of intelligibles, that is, a world composed not of [purely] physical objects, but of occurrences which have meanings . . . occurrences in some manner recognized, identified, understood and responded to in terms of this understanding.

Here, even when we encounter something that appears meaningless to us, and perhaps we say that we can make no sense of it, this experience can only occur against and within a background world of meanings. The "meaningless" thing occurs precisely *as* something that *lacks meaning*, thereby reinforcing both that it is so situated, and that tacitly it has meaning – as the very designations "something" or "it" convey. In such experience, we are particularly impressed with a meaning yet to be discovered and the thing exists for us – that is to say, its present meaning precisely consists in – the respects in which it is felt to lack meaning.

Clearly, such a space in which things encountered always already have some kind of sense through being embedded in a world of interwoven significances is not to be equated with – i.e. conceived exclusively in terms of – the space of "purely physical" scientific/geographical locations. For example, geographical locations can have particular personal or cultural significances that greet us as we enter them. Awareness of these can be a vital constituent of being more fully present in the place. But they are not the only sources of significance present in the encounter. For example, one can, as it were, to a considerable – perhaps very high – degree take one's personal significances with one as one travels from one geographical location to another. For example, what I might come to identify as certain core beliefs and what have become habitual ways of construing things need not change – can be carried forward – as I move from one place to another. The degree to which this is so in part will depend on the power of the new place, the manner of my travel, and my susceptibility to change. In many circumstances, experiencing changes in one's outlook as a result of new encounters can be regarded as an entirely healthy, and refreshing, de-centring of self. But, for the purpose of revealing the extent of the tacit play of anticipation in our inherence in the world, consider the following case. Suppose one were to be completely removed from the milieu of those things/people in which one customarily has one's being, enjoys relationships, might not something in excess of a healthy de-centring occur – something more akin to an "unselving" that can be seen to parallel the "unselving" of things in nature discussed in the previous chapter?

To take an extreme case: suppose that someone is forcibly removed from the family, companions, computer, books, music, furniture, vistas, and so forth that, as customarily experienced and located (at least in part) constitute their home. They are no longer able to participate in its daily routines and associations, to enjoy, say – and taking one small aspect – the familiar and yet always surprising garden views across the seasons, the blossom and shade of its trees, the demands for weeding, its promptings as a historical site of personal events, aspirations, and imaginings. Necessary as certain circumstances might seem to make it, a child's or an old person's being extracted from the milieu that this exemplifies and "taken into care" might not be without its cost in terms of sense of self. Conversely, it might be argued that finding oneself in a strange environment might heighten one's sense of self – one becomes more self-conscious in negotiating unfamiliar artefacts and expectations, one feels more sharply defined against the surrounding other that is now unfamiliar rather than familiar. For the moment, I leave this as a possibility, for first I would like to explore further this notion of "promptings" and its significance for understanding the relationship between the individual and the environment.

There is an obvious sense in which different places/situations/neighbourhoods prompt – call for – different responses: whether we are in the presence of a friend or a stranger, are at a birth or a burial, encounter a valley or a mountain. This is true both in general cultural terms and at a personal level and on many occasions it is taken to depend on what are referred to as "associations" that the place or occasion might have. But this way of putting the matter is already misleading in some cases. Such talk of associations sets things up in a way that suggests that this place exists prior to them, had some independent existence to which the associations subsequently became attached. While when viewed objectively things can appear this way, such a perception can be the reverse of personal experience. For Jane, say, this *is* the place where her husband is buried; the stone and the grass and the faded flowers speak of it through and through. This place *claims* her in this way – and in so doing it becomes part constitutive of who she is. She *is* the person who is so claimed and to be other would require an act of severance on her part. Her being as the individual that she is and the being of the place in its uniqueness in experience are intimately interwoven. To refer back to the language of the first part of this chapter, through their emotive power particular emplaced intentional objects of her consciousness have become definitional of her.

It is true that this is an extreme case and has been simplified, but it is illustrative of a general phenomenon that is central to our emplaced intentionality. We are *always* claimed by places. With a seeming exception to be discussed presently, phenomenologically we are never unplaced – though our sense of this is often highly tacit and the claims that are made upon us vary greatly in quality and strength. But no human action is possible in the absence of place, nor, indeed, any thought, including that of sense of self. All that we do and think involves a sense of place and arise amidst a milieu of prompts and claims that are experienced as emanating from this place. Consistent with the intentionality of consciousness elaborated earlier, what this amounts to is that one *is* in one's dealings with a world – partly familiar, partly unfamiliar – that is taken as "not me", "other", in a fundamental sense.

But it is important to note here that although experienced as other in this sense, this world that we inhabit in its otherness nevertheless must be *self-assuring* in a primal sense of being at some level and to some degree receptive to the self. We exist through participating in an intimate, free, and frequently non-verbal – often bodily – dialogue with it, as, in its occurring, it provides, as it were, the questions to our replies as well as the replies to our questions. For example, on a woodland walk, we might encounter a fork in the path and instinctively choose the right-hand way, while on another occasion we might seek some sign as to which way to go – perhaps one way promises to be less muddy than the other. Here, a *mutual anticipation* (and invitation) of self and world is in play. The path in its otherness and mystery awaits us and receives us as we walk, just as we await and receive what it offers. In this self-assuring interplay, each – the walker and the path – is called into being, into occurring.

The significance of this account for the present argument is that it shows that there is a level at which self-assurance and intelligibility are inextricably interwoven. In addition, an affective aspect of being in a world is acknowledged that includes our bodily apprehension of where we are. There is a sense – i.e. feeling – that however hostile a situation might be, some aspects at least mesh with our anticipatory structure, that is to say, they are what we took them to be. Without this minimum level of felt mutuality, the situation could not even be experienced as hostile because there is no foothold for understanding it at all. One would have entered, as it were, ontological freefall. This returns us to the seeming exception mentioned previously: the case where a place has no self-assuring features, that is to say *nothing* is as we took it to be and therefore it lacks any intelligibility whatsoever. Here we would be truly adrift; literally, lost – our intentionality, one might say dissipated in blank space. With the complete confounding of anticipation comes the complete dissolution of both place and self. An unknown or mysterious place may breed anxiety, but the fully "no-place" breeds no-thing and no-self. In sum, some things must be as they seem and be capable of claiming our attention, posing questions, intimating what kinds of responses would or would not be fitting. This is a minimum requirement for being *there* in a place, for a basic level of intelligibility to be present. Here is a key sense in which our transcendence is always emplaced.

Viewed as a matter of degree, the idea of a pathological unselving can become salient for educational institutions in a number of ways, some of which are of particular significance when it comes to learning in and from nature. For example, insofar as formal schooling is organized around a pre-specified curriculum, this necessarily tends to insulate learning from the free play of mutual anticipation. It therefore inclines teachers to an indifference to those ways of being in the world and of sense-making that spring from pupils' own emplaced experiences. This both drains the intelligibility of the things that they encounter and learn of an important aspect of its vigour, and neutralizes the felt vitality of the proximate environment such that the free dialogue to which allusion was previously made cannot be sustained. The vibrant occurring of things in nature becomes heavily cabined or altogether effaced. This being so, I will now discuss some further nuances present in the play of emplaced anticipation in our experience and sense of selfhood, and the ways in which they situate us in nature.

The play of anticipation

As previously noted, perhaps few today would argue for the pre-existing fully autonomous, disengaged, self of classical rationalism (and perhaps liberalism?) even as a theoretical construct. Nonetheless, it remains a useful contrasting idea for differentiating more relational notions of the self. One such commonly experienced relational aspect is that of the importance of geographical and cultural origins to many people. While such an emphasis can bring dangers of a philosophy of "blood and soil", ideas of "rootedness" in a culture and/or place as constituents of self (identity) are frequently encountered. Environmental psychologists make the point that selfhood is seriously conditioned by sense of place and hence that human beings are fundamentally geographical beings. Not only, as some have it, can you not "take the country out of the person", if you take the person out of the country, somehow their ability to *be* themselves can be reduced. The necessary possibility of mutual anticipation of self and world can be eroded.

The claim that the disruption of such anticipation could be emancipatory through provoking new kinds of receptiveness, sensitivity, raises the issue of what reference to the idea of anticipation invites. For example, if anticipation were interpreted as highly routinized, it would indeed seem to exclude or diminish the element of receptivity. And some uses of the term in everyday discourse suggest this as when, for example, a hotel recommends itself as anticipating the needs of its guests by providing a set of facilities and procedures that are on hand to support and reinforce a particular pre-specified lifestyle. But there are other senses that are more germane to the present argument. For example, the anticipation experienced on a fine spring morning by the walker as she sets off, or of the fisherman as he approaches the riverbank at dawn, or that of the trysting lover, is of a very different calibre. Here anticipation is experienced as an openness to and embracing of the unknown that is to come – the challenges and the sights, the smells, the textures, the ambiences, and surprises of, say, different places and times of day. It speaks of a keen attentiveness. Such anticipation quickens life, gives a heightened sense of being. It is a form of futurity and of true ecstasis.

Also, there is a sense in which everything we do involves anticipation: the gardener that the soil will yield to her spade, the sitter that the chair will bear his weight, the walker that the Earth will bear her up, the reader that the text has a meaning. Usually these are not conscious expectations, but are thoroughly implicit in the very movements/behaviour of the limbs of the gardener and walker, in the very act of scanning the text by the reader – indeed, in the act of opening the book or envelope. Anticipation pervades our being at many levels and it lies at the heart of the constant delicate, intelligent, adjustments that we make within our environment – the aforementioned examples are intended to indicate that the intelligence involved here is often bodily. Hence, when I speak of the self or environmental consciousness, this should not be equated with traditional (for example, Cartesian) notions of the mind as essentially separate from the body: some sort of mental entity in which discursive thinking occurs. In terms of this sort of lexicon, I would be positing an indissoluble consciousness of "mind–body". For this reason, I will sometimes use terms of this kind, and also speak of embodied or corporeal consciousness, the point being that what I take to be of ultimate concern is that of which to some degree either we are conscious or are capable of becoming conscious. This would include all sensory perception, including awareness of inner bodily sensations of whatever kind (kinaesthetic, soma-aesthetic, etc.). It follows that by "bodily" I do not refer to the purely physical and by "frame of mind" I do not refer to the purely mental (in traditional terms).

Returning to the significance of the term "anticipation", it might be asked: why speak of mutual anticipations rather than simply speaking of significances? It seems to me that there is a sense in which the latter can be construed as achievements (however fluctuating or ephemeral), whereas the former are to be thought of more as processes of directedness – understandings, yes, but *understandings in motion*, ever uncompleted aspects of the flow of intentionality. Again, it might be thought that the notion of "expectations" comes close to this and, indeed, it is a frequently used term in some of the contexts considered previously. But it seems to me that this denotes something perhaps that is still too abstract and explicitly statable – too cerebral – whereas anticipation is consistent with the greater immediacy that is characteristic of much of our (mind–body) intercourse with our proximate environment. As understanding in motion, it is in constant fluid readjustment. For example, should we trip, it seems to make more sense to say that the body anticipates a fall (by, say, rapidly shaping itself to take the impact) rather than that it expects such. In sum, I use the term "anticipation" to identify something that is central to the openness of environmental consciousness: it is its anticipation of otherness that constitutes such consciousness as an emplaced transcendence that is each individual's own receptiveness to the world – *its* kind of waiting upon, its own attentiveness to, the other. Harking back to previous discussion, individual consciousness is not simply a blank slate that the other "writes upon" or "subjectivates", but reaches out from itself in conditioned openness, through its own anticipations, towards the other – just as the other reaches out towards it in response and "fulfils" these anticipations in its own ways, often provoking modified or new anticipations.

It follows, and it is important to emphasize here, that anticipation as I have been discussing it is not to be regarded as simply our projection onto an inert world. It occurs in the context of our participation in places, and, as previously noted, a place and the things that populate and constitute it can be experienced as awaiting us and as claiming us through the invitations and prompts that they offer. To consolidate this, perhaps we look into the kitchen and see the dirty dishes as awaiting our attention, the shade of a tree beckons us on a hot day. We can experience the history or ambience of a particular place as deeply affecting our sense of who we are and what we are doing. This is true whether it is our home with its familiar utensils that anticipate and invite our activity there, our workplace, or the endless lines of white war graves of the Somme that silently await our coming and remembrance. The significances that we experience in these examples are not merely subjective additions to something more primal or real, rather they actively constitute the world of sense in which we live – and from which all else is abstraction and fabrication.

Furthermore – and equally important – previous discussion has made it clear that this sense of reciprocal anticipation in our experience of the world emanates not only from the artefactual, as exemplified earlier, but also from the quintessentially non-artefactual: nature. Mention was made of the spider's web anticipating the stray fly, the swelling buds standing out on dark stems anticipating warmer and longer days, and so forth. These are examples of anticipation in play "within" nature, but the boundary between nature and us is highly permeable: reaching for a ripe fruit, it can be experienced as awaiting our grasp; the nearby robin awaits alert for the disturbance that will expose hidden grubs as I take my fork to the soil. An important part of our connection with the natural world is facilitated through our participation in this interplay of anticipation. Without it, indeed, we would (again) enter ontological freefall, for, as previously argued, there is an important sense in which we ourselves inhere in the world through a meshing of our anticipations with those experienced in the places in which we live and in which essentially we find ourselves. In the case of "natural" places, this can range from an unreflective picking of an inviting apple to a more general attunement to the myriad signs of what is nascent on a spring day that shapes our own anticipations and thus locates us in the play of the seasons.

Ultimately there is nothing purely objective or passive about a place – domestic or natural; it only appears so when we have lost touch with its, and our own, genius – as when, under the influence of scientism, we can be persuaded that to recognize its transcendent, inviting otherness is to indulge a frothy fiction. Hence, it is perfectly legitimate to speak of the shade of a tree beside the cool pool as beckoning us on a hot day, the wayside rocky outcrop as inviting us to sit after a long climb. A mountain can be experienced as challenging us to attempt an ascent or as deterring us in our path. We speak of the view that awaits us over the brow of the hill. In the act of stretching out our hand to the ripe fruit, it can be experienced as anticipating our touch and as waiting to be picked. In approaching land after a sea trip, we can experience the land as also approaching us, and its terrain anticipates the visitor as perhaps brooding or as welcoming. To experience terrain as hostile or friendly is to participate in mutual anticipation: the terrain is felt as hostile towards us and it is experienced as in an active relationship with us, expressing an agency. This is all part of the character of our emplacement.

Articulating nature

Reasserting this dimension of our relationship with the natural world is highly relevant to ideas of ecologizing education, and, indeed, in one respect at least, returns us to an issue that lies at the heart of any such ambition: the kind of language that it is appropriate to use in describing nature and our relationship with it. Part of the reason for confronting scientism throughout this book is because it proscribes descriptions of the kind given earlier. It will object to the idea that something taken to be inert or lacking sentience, such as a mountain, literally can challenge anyone, or that a view or a ripe fruit can await us. Things of this kind are just *there*, and no more, and to suppose otherwise is simply to indulge some romantic illusion. It is perhaps now time to respond more fully to this line of argument than has been done so far. There are a number of strands to this response that can be articulated in the following questions:

> Are scientific descriptions as accurate and truthful as the objection assumes?
>
> What does adopting scientific descriptions to the exclusion of all others – or, setting them up as superior to all others – do to our understanding of the world in general and nature in particular?
>
> Does this enhance our relationship with nature?
>
> Does it better prepare us to address the threat of environmental catastrophe that now confronts us?

Let me begin by saying something more about what I mean by "scientific descriptions" in relation to the natural world. Put simply, here I refer to descriptions that are couched in terms of purely physical phenomena that in principle are publicly observable or can be inferred from such. As previously noted, these observable phenomena and their relationships are understood to be ultimately material: constituted by fundamental particles and forces, that increasingly are taken to be fully accountable in mathematical terms. Hence, the occurrence of a particular colour such as blue is taken to be more accurately described as being a colour of a particular wavelength. These descriptions are understood to be objective in the sense that they are the product of impartial observation, purified of any emotional attachment or sentiment. They are, in the vernacular, the "hard facts" that are available to any reasonable person who suspends all personal bias and emotion. It is not that, say, scientists do not feel emotions such as excitement and wonder (and, no doubt many more) during the course of their work, rather it is that their scientific accounts are not swayed by these in the sense that these accounts do not go beyond what can be objectively observed or inferred in the way described.

Now, undoubtedly, this way of approaching the natural world has brought many benefits and has been highly successful in its own terms – for example, few would wish to be without the benefits of potable water, antibiotics, and anaesthetics. Clearly, being able to categorize, theorize, and predict phenomena has an important role to play in understanding the natural world and in making decisions about how to respond to it and to events that affect it. Knowing that increased amounts of gases such as carbon dioxide and methane in the atmosphere that are the product of human activity contribute to global heating, and gathering data on what the effects of this heating are likely to be on the climate, are invaluable in making judgements about how to respond to the dangers of climate change. But with regard to what our basic orientation to the natural world should be, this kind of knowledge is inadequate. And it is the character of this orientation that ultimately is fundamental to what our responses should be to environmental problems. To give a hypothetical (and much simplified) example: suppose technological advances made it possible to sequester our greenhouse gas emissions and to deflect the sun's rays to the extent required to halt global warming, but with collateral effects that resulted in the denaturalization of large tracts of land and the disruption of natural systems. How should this course of action be judged against an alternative of, say, making extensive changes to our energy, economic, and lifestyle requirements? Here arise issues concerning the intrinsic integrity and value of nature and their relationship to our well-being – including what we should *mean* by human well-being. To address them, we require a knowledge that has been effaced by the scientific approach: a knowledge of the being of nature in its occurring that has been the subject of this and previous chapters and that for its articulation requires a language that far out runs the language(s) of the natural sciences.

But, now, in saying this, is there not a danger of seeming to legitimate the possibility of language of almost any kind as a true expression of nature? If we abandon the criterion of language needing to express the "hard facts", what is there to stay, for example, sheer anthropomorphism – or even seemingly sometimes random attributions of human thought and emotion to things in nature in the way that can be found in superstition or fairy tales? The answer to this lies in the quality of our openness to things in nature. Certainly, the metaphysics of mastery is not the only source of distortion here; unwarranted sentiment and laziness of perception are large dangers. Perhaps the depiction of animals in young children's literature and films as having human sensibility (perhaps, dressed in human clothing) is harmless. Perhaps it initiates some kind of simpatico with nature that can be the germ of a more grounded pathetic connection in later life. But this tactic must be transitory. True environmental consciousness aspires to be true to the occurring of things in their being as they present themselves. It is there, with them, participating in them, the overweening instrumental and stereotyped preoccupations of everyday consciousness suspended ("bracketed" as Husserl might say). It attempts to be true to the intuitions that arise under these conditions. More will be said on the veracity of these intuitions in Chapter 6 in which the idea of nature's possessing a voice is explored. But at this stage, it can be confirmed that the form of eco-phenomenology being advocated in this book seeks to combat the disenchantment of nature that scientism has brought in its train by refusing to proscribe the languages of purpose, value, and emotion in the articulation of our experience and understanding of the natural world. Its underlying stance is realist but not materialist.

A question of time

There is one further – and highly germane – way in which environmental consciousness repudiates the strictures of scientism when it is engaged with the natural world: the kind of time within which it participates. In Chapter 2, the point was made that lived natural space – that is the space that is vouchsafed when we participate in the native occurring of things – is not that of uniform Euclidian space but rather one of unique locales and neighbourhoods. Hence, primordially it is not susceptible of being measured or determined through standard units such as metres, miles, or light years. So, too, phenomenologically, natural time is not to be conceived as uniform and capable of being gauged in terms of standards units. It, too, occurs with the occurring of things and its quality and duration reflect this. It is not the case that first of all there is some neutral universal time framework into which subsequently things are located and accounted for. This kind of abstract conception is a product of a form of discursive thinking – an ordering projection upon reality. Primordially – phenomenologically – *first of all there are emplaced things that*, as it were, *in their native occurring bring time with them and initiate its many experienced qualities*. Primordial time is pregnant with meaning. There are times of hope, wonder, anticipation, foreboding, peace, restlessness, and so forth – all intricately nuanced in their occurring.

Full participation in the native occurring of things in nature is oblivious to chronological or "clock" time. The superimposition of mechanized clock time upon what hitherto had been experienced in terms of fluid natural phenomena such as the passages of the sun and stars across the sky, night and day, the manifestations of the seasons, and the internal sense of one's own growth and decline, transforms the experience of time. Drawing on David Landes' *Revolution in Time: Clocks and the Making of the Modern World*, Amy Shuffleton (2017) discusses how clocks were introduced in Europe to synchronize human actions in the public spaces of churches and workshops, eventually enabling modern factory production and transportation networks. They became central to ideas of social efficiency and the scientific management of labour (including that of educators – here she cites language that speaks of loss of "instructional minutes" that arose in debate about the use of standardized achievement tests in Chicago public schools).

The division of time into precise standard units represents a radical instrumentalizing of time, and, from the perspective developed in this book, an attempt at the mastery of time: it is brought into service as a measurable and measuring commodity, transmuted into something that can alienate and control, along with everything else when held in the sway of a metaphysics of mastery. Everything becomes a matter of pre-specified future achievements. As such, the unfettered generalization of mechanical time into public consciousness is a deadly assault on an environmental consciousness that seeks to be open to the primordial reality of things in nature. Indeed, such environmental consciousness may be heedless of linear time altogether. While the experience of anticipation as discussed previously implies a sensing of a kind of futurity in our experience of the arising of things in nature, this is not such as to be measured in standard units and nor is it strictly linear in the sense of a movement from some briefly existent present punctuated from some yet to be future. Rather it is experienced as an extension of an all-embracing present. From a chronological perspective, this present, this being present, is timeless – the immediate experiences of the native occurring of things in nature as illustrated in Chapter 2 are had in timeless moments. Everything is here and now. Even the sense of spiral or cyclical time that nature expresses when we stand back and contemplate its processes can be held in abeyance, or occur not as something exterior to, but as inherently part of, the present.

Here arises a sense in which there exists a tension with conventional ideas of education. Education involves change, a moving forward from one state to what is considered to be a better state. It is energized by some notion of a future achievement – one that in the late-modern period is increasingly pre-specified and strident. It involves what Karsten Kenklies (2020) has termed a "dissociative disorder", in that it requires a constant minimization of the present in the interests of progressing to a more desirable future. Here a pathological imbalance reigns within temporality between present and future, and between temporality and atemporality. One is constantly distracted from dwelling in the present moment – alienated from ones being *there* by an overbearing call to some future achievement. Indeed, from the perspective of much contemporary educational thinking, to dwell in one's present state innocent of any future ambition is considered a waste of time, a form of idleness, laziness. Yet, if previous argument is accepted, it is precisely in such moments of giving oneself over to the sheer occurring of things, their immediate living presence, that consciousness is at its purist and most authentically engaged in the play of being and the occurring of truth. Here we have a glimpse of another kind of education – one freed from the metaphysics of mastery, and that will be explored further in Chapter 7.

For the present, it has to be granted that these timeless moments cannot last indefinitely. For many, in life as we currently have it, the demands of the everyday world will press in – perhaps, will be felt with increasing urgency. Yet maybe something of value can continue to flow from such experience of original thinking. Perhaps the reverberations of that experience sometimes can be sufficient to disturb and revitalize more routine thinking, re-sensitizing it to what lies beyond its current purview. In this way intimations of new ways of relating to the world might emerge. Although now to a degree withheld, the play of being might yet come to echo in how we comport ourselves in everyday life.

Notes

1 It should be made clear here that two senses of "intentional" come into play: first, a direct pre-predicative engagement that is to be distinguished from; second, having an intention in the further deliberative sense of, say, working on a problem or deciding a course of action. Each can feed into the other. In what follows, the reference is mainly to the pre-predicative sense of intention, although it is part of the broader picture that such pre-predicative engagement plays into intention in the more deliberative sense.

2 It is true that in his desire to make philosophic contemplation impersonal and dispassionate as free as possible from human hopes and fears, customary beliefs and traditional prejudices – Russell claims that the free intellect "will value more the abstract and universal knowledge into which the accidents of private history do not enter, than the knowledge brought by the senses, and dependent, as such knowledge must be, upon an exclusive and personal point of view and a body whose sense-organs distort as much as they reveal" (p. 93). This discounting of the role of affect and the body in sense-making clearly goes against some central themes of this book. But his cabined notion of understanding does not vitiate the point that from within the constraints of a very different viewpoint emerges the essential underlying insight that there is an intimate relationship between enlargement of self and openness to the otherness of things.

3 See the distinction drawn between the metaphysics of "objects" and the metaphysics of "things" in Chapter 2.

5

ANTHROPOCENTRISM, ECOLOGICAL JUSTICE, AND POPULATION GROWTH

In this chapter, a recap of the ways in which anthropocentrism and the metaphysics of mastery condition thinking about nature and the environment is illustrated by reference to the politically charged issue of human population growth and the influence of an underlying principle of social justice. This invites a consideration of ideas of *ecological* justice that reject current rampant "human supremacism" in the distribution of the Earth's resources. Here, the issue of nature's intrinsic value arises anew and some philosophically influential versions of this – and the ways in which they rebut anthropocentrism – are critically evaluated. This leads to some refinement of ideas of nature's intrinsic value elaborated in earlier chapters and to revisiting ideas of emplacement and normativity in nature. There follows a discussion of the principle of social justice in the light of these considerations.

Anthropocentrism

Some 20 years ago (Bonnett, 1997), I used the term "great divide" to express the character of the relationship between humans and the rest of nature as portrayed in many prevailing aspects of Western culture. I referred to the several ways in which influential cultural-religious-scientific assumptions and proclamations incline us to view ourselves as separate from, and superior to, the rest of nature – the latter having significance and value only insofar as, and in the ways in which, we choose to bestow them. Under the (often tacit) influence of these powerful views, nature appears essentially as something to be overcome and as a resource: there to be explored, interrogated, tamed, and manipulated in the furtherance of our self-given purposes. In other words, our attitude is one of extreme anthropocentrism. Nature is taken to have no truly intrinsic value, only instrumental value: we use it; it gives us pleasure. Along with others, I have attributed this attitude to the modernist humanism that arose from the European Enlightenment (Bonnett, 2015b) and in Chapter 3, I elaborated it as a "metaphysics of mastery" because it installs us in a version of reality in which the essential character of everything that appears is that of help or hindrance to the exercise of our will.

As previously discussed, this frame of mind has many important consequences when it comes to identifying and addressing environmental issues. Importantly, it also conditions our views on principles that are of very broad educational significance – ones that in turn inform how we frame and are willing to discuss potential solutions to these environmental issues. In what follows, I examine how this plays out with respect to the principle of justice – in particular in relation to a particularly pressing and politically sensitive issue that environmental concern raises: population growth. If nature has intrinsic value, the impact of human population growth upon its well-being becomes a moral matter. I begin with some preliminary discussion of this.

Human population growth

Notwithstanding an apparent drop in the global fertility rate over recent years, a global human population that stood at 7.6 billion in 2017 is projected to reach 9.8 billion by 2050 (UN, 2017). It has also been estimated that the ecological carrying capacity of the Earth is between one and two billion (Daily *et al.*, 1994).[1] To be sure, these figures are open to question on many counts. Clearly, long-term estimates can be subject to all manner of future contingencies such as the impact of new technologies and unforeseen natural events, and the criteria for sustainable ecological carrying capacity need clarification and examination. Nonetheless, the aforementioned estimates are widely cited and in very broad terms they give an indication of the scale of the problem that we are facing. In addition, they express an unacknowledged but important philosophical and ethical predisposition: a continuing underlying anthropocentric orientation whose focus is on ecological carrying capacity of the planet in terms of maximum *human* population. Framed thus, the implicit invitation continues to be to perceive the natural world in terms of its ability to service the material needs of an ever-growing humanity. While for many, this focus on human needs might be entirely uncontroversial, presently, I hope to show that there are good reasons for not simply assuming this position. In addition, it is worth noting that there is a growing literature that challenges the assumption that the satisfaction of human desires trumps all else. For example, David Foreman, founder of the "Earth First" movement, has argued that they should be subordinated to the well-being of the biosphere (Foreman, 1993), and more recently the idea that "nature needs half" (of the Earth's resources) has attracted the support of some writers prominent in the field of environmental literature, such as Noss and Cooperrider (1994), O. E. Wilson (2016), Helen Kopnina, and Haydn Washington (2018). I will turn to some of the ideas that are expressed in these views shortly.

First, let me return to the idea of human population control. Simply to raise the question of the need to control the human population if sustainability is to be achieved can elicit fierce criticism. As these criticisms can illustrate a number of powerful assumptions and issues fundamental to developing a defensible position on how humankind should behave towards itself and towards nature – including what a well-informed principle of justice would demand – I list some of the key arguments. These have been usefully gathered in two recently published papers (Kopnina, 2016; Kopnina and Washington, 2016). Those who argue that a burgeoning human population is both a hindrance to achieving sustainability and to protecting non-human nature are criticized on the following counts:

1 The view smacks of Western elitism and hypocrisy: those whose history and current lifestyle have been the chief cause of ecological problems look to the most vulnerable human communities who leave the smallest "footprint" to restrict their material development and utilization of natural resources. Hence, calls from those living in economically developed nations for conservation and for elevating the moral status of nature are specious because in practice others will bear the main burden of this stance;
2 It turns attention away from key issues of social justice and more equitable distribution of wealth: if the West/North consumed less, there would be enough for everyone;
3 It ignores the deleterious role of neoliberal free-market capitalism that is complicit in ever-increasing economic/material growth and therefore demands on natural systems;
4 It expresses an anti-human stance that feeds off a Western highly romanticized myth of nature;
5 It ignores the practical irrelevance of population growth to achieving sustainability when the aforementioned points are taken properly into account.

Sentiments of this kind have historical antecedents. The revival of anxiety over a Malthusian "population explosion" that arose in the mid-twentieth century was dismissed as "reactionary nonsense" espoused by a "comfortable elite who wish to maintain their status" by the historian J. D. Bernal in his then influential *Science in History* (1969, p. 973). He held that population growth was to be welcomed and that it easily could be sustained by the growth of applied science, claiming, for example, that under the then Soviet regime: "Already they are working on the scale of Nature. Geography can no longer be taken for granted, the world surface will henceforth be what man chooses to make of it" (p. 968). Mechanization will enable, for example, "the fullest use both physically and chemically . . . of every drop of water. Not only the great rivers but every little tributary is pressed into service" (p. 967). Under socialism, there is the potential for "an enormous extension of civilization – agriculture and industry – together – in which the soil will not merely be preserved but indefinitely improved, and the life it supports will be multiplied" (p. 973).

Overall, it can be seen that a significant part of what lies at heart of these criticisms of human population control is both a superordinate idea of social justice and a repudiation of the notion of a transcendent nature that warrants moral recognition (not to mention, in the case of Bernal's view, a rampant metaphysics of mastery of the kind elucidated in Chapter 3). It is the business of this chapter to explore some of the synergies and tensions between this orientation and powerful notions of *ecological* justice that have arisen in the literature of environmental concern. It will be held that some versions of the latter have radical implications for how we think about justice more generally – and, indeed, the whole of moral education.

Ecological justice

I begin with a notion of ecological justice that is closely aligned with mainstream ideas of social justice. It draws attention to a need to extend the application of social justice into areas that previously have received too little attention. A good example of this kind of view is the "eco-justice pedagogy" that has been advocated by C. S. Bowers (2002). This has three main foci:

1 To develop an awareness of the environmental racism and class discrimination involved in the way that the deleterious environmental impacts fall disproportionately on ethnically and economically marginalized groups;
2 The recovery of non-commodified aspects of community through a reversal of an ever-increasing dependency on meeting life's daily needs through consumerism rather than through self-reliance within the family and within networks of mutual support within communities;
3 To develop a sense of responsibility towards future generations and a corresponding self-limitation by an expansion of non-consumptive relationships and opportunities to develop personal talents that enrich the community culturally (as against materially).

Bowers claims that central to this enterprise is the identification of root metaphors embedded in language that shape the way that we engage with the world: currently, for him, metaphors that underlay the industrial revolution, such as "individual" and "linear progress". He argues that these systematically undermine the value of tradition and therefore the inter-generational knowledge and continuity that are necessary for ecological wisdom (and that are so central to indigenous perspectives). Also, these modernist metaphors conflict with the root metaphor underlying an eco-justice pedagogy which is "ecology", and that "foregrounds the relational and dependent nature of our existence as cultural and biological beings".

Here, briefly sketched, is a view of ecological justice that can be seen as enriching mainstream views of justice, drawing attention both to its environmental dimensions and particular ways in which language can be complicit in undermining it. In this view it would be a task of education to enable students to identify and critically evaluate such sociocultural patterns in the hope of questioning consumer dependency and an acceptance of environmental degradation as a necessary trade-off for achieving personal convenience and material success. As portrayed, while enlarging the scope of social justice, this view of ecological justice requires no modification of its core meaning (whatever, precisely, that is taken to be[2]) and presents no overt connections with ideas of limiting population growth. It remains fully compatible with the anthropocentrism inherent in traditional versions of the principle of social justice. The challenge to social justice as a fundamental concern arises with a different view of ecological justice: one that requires respect and a fair distribution of resources not only for humankind, but also for the natural world.

But what is the basis for attributing moral standing to non-human nature, for claiming that it should be respected not simply for its capacity to satisfy human demands, but for its own sake? In what sense does it possess inherent intrinsic value?[3] Such questions lie at the heart of the ambition to develop a non-anthropocentric notion of ecological justice that embraces the whole of the biosphere and that, for example, is concerned with a fair distribution of resources between human and non-human nature in a situation in which the limitations of such resources in meeting both the ever-growing demands of humankind and sustaining many aspects of non-human nature are only too apparent. For example, because of human appropriation of the natural world, in less than two human generations the population of vertebrates has fallen by a half, with a number of apex predator species – but not only these – on the verge of extinction through habitat destruction and hunting (WWF Living Planet Index, 2014).

Previously, I alluded to the "nature needs half" movement. Kopnina *et al.* (2018) identify two principles that underpin this idea:

1. All species have a right to continued existence. It is morally wrong for human beings to cause the extinction of other organisms;
2. Habitat loss and degradation today is a leading cause of the decline of biodiversity.

Looking towards some perhaps not so distant future (given the current rate of mass extinctions), even if through advanced technology and management techniques it were possible for a "post-ecological" human civilization to exist upon a thoroughly denatured planet, it is held that, morally, this is not an acceptable prospect. As Eileen Crist (2012) puts it:

> What is deeply repugnant about such a civilization is not its potential for self-annihilation, but its totalitarian conversion of the natural world into a domain of resources to serve a human supremacist way of life, and the consequent destruction of all the intrinsic wealth of its natural places, beings, and elements.

Central to this rejection of "human supremacism" is the idea that nature – or at least some aspects of it – has intrinsic value. But how, precisely, is this to be understood? And given the dominance of anthropocentric assumptions in mainstream ethics and accounts of justice, how is it to be defended?

There have been numerous articulations of the idea that nature possesses an integrity and intrinsic value that warrants moral regard. I have developed one such argument in Chapter 2 based on a phenomenological perspective. As a preliminary to amplifying this argument and augmenting a defensible position on this fundamental issue, I outline in the following some key components and alternative arguments. These seek systematically to reveal and erode what is taken to be an extensive and often unacknowledged anthropocentrism in our conceptualizations of the world in which, in reality and in many respects, we are embedded as just one participant amongst others. It is held that this anthropocentrism supports a hubris in our perception of the natural world that obstructs a truthful engagement with it. Views of this kind point to a colonization of nature that goes deep and it will take a great deal to dislodge it so as to allow the marginalized (nature) a voice. This being the case, it is to be expected that initially, when viewed against this deeply ingrained cognitive orthodoxy, attempts to clear a space for this voice to be heard and articulated can appear outlandish, or even fanciful.

Nature's intrinsic value (the rejection of anthropocentrism)

Historically, there has been a wide range of views that celebrate nature's integrity and in some sense ascribe value to it: for example – and in very broad terms – Daoism and Buddhism in the East and Romanticism and New England Transcendentalism in the West. In their different ways, each has elevated the idea of nature such that reflection upon its intrinsically vital character, and our place within it, come to the forefront of understanding ourselves and our flourishing. To the extent that it can be shown that human values are formed and informed by contact with nature, there is an important sense in which nature acquires the role of teacher of human values. It ceases to be a mere satisfier of pre-formed (and frequently consumptive) values and becomes an important source of inspiration in value formation. Yet this is not of itself sufficient to substantiate a claim for intrinsic value in nature, for its value remains viewed in terms of serving human ends, even though, as now conceived, it can be instrumental in shaping these ends. In what follows, I refer to a sample of recent views that illustrate the more ambitious theme of ascribing intrinsic worth to nature in the sense that just as humans must be regarded as ends in themselves, so, too, must nature. Insofar as this is true, the unbridgeable moral divide between humankind and nature promoted by Enlightenment thinking is further eroded: nature should command a commensurate respect for its intrinsic value as does humanity. I begin with a particularly well-thought through exposition of this position given by Paul Taylor.

Taylor (1986) argues that there is a fundamental equality between all species of animals and plants on the grounds that each is a teleological centre of life and that

> [t]o say it is a teleological center of life is to say that its internal functioning as well as its external activities are goal-orientated, having the constant tendency to maintain the organism's existence through time and to enable it successfully to perform those biological operations whereby it reproduces its kind and continually adapts to changing environmental events and conditions. It is the coherence and unity of these functions of an organism, all directed towards the realization of its good, that make it one teleological centre of activity.

An important feature of Taylor's view is that being a teleological centre of life does not require organisms to have consciousness – only that "they have a good of their own around which their behaviour is organized". He goes on to argue that:

> Under this conception of individual living things, each is seen to have a single, unique point of view . . . determined by the organism's particular way of responding to its environment, interacting with other individual organisms, and undergoing the regular, lawlike transformations of the various stages of its species-specific life cycle.

Unlike the view of Jay McDaniel to be described shortly, Taylor sees this as applying only to plants and animals, each of which can be conceived as having a good of its own through being a centre of goal-orientated activity – *its* goals in terms of which we can give teleological explanations of why it does what it does. Claims of this kind are not uncommon. For example, we find something very similar in the views of Holmes Rolston III (1997b) who speaks of living organisms as axiological systems, each having and defending a "good of its kind". Indeed, the view that there is an important sense in which the lives of plants and animals are goal-orientated seems quite widespread, for example, as previously noted in Chapter 3, it is implicit in the language we often use to describe their activities – the predator hunts its prey, the parent protects its young, the roots seek moisture, etc.

Given the central concerns of this chapter, two questions arise in relation to arguments of the kind outlined above: (a) how literal should we take them to be – particularly in the context of identifying moral equality with human beings? (b) even if it were agreed that plants and animals were to be properly viewed as axiological, why should we attend to this, value it? How does it follow from the claim that an organism is goal orientated that we should respect its goals? Is it wrong to seek to eradicate malaria and Ebola? These are questions to which I will return. First, I refer to a recent view that connects concerns for sentient nature with broader social issues.

In a recent paper, Peter Kemp (2017) poses the following question: does good citizenship demand the care of animals and plants? (particularly at a time when biotechnology allows us to experiment on animals and use them as commodities). He argues that:

> We must recognize that humans and animals share a corporal life so that neither can be reduced to things, but rather we must confer to each both sensitivity and feelings (including the capacity to suffer), and we cannot deny that most developed animals have a consciousness of a surrounding environment and even, in particular situations, an intuition of their imminent death.

However, according to Kemp, although animals are "ends in themselves", only humankind possesses a sense of moral responsibility:

> It is true that the human responsibility for animals presupposes that we humans are capable of identifying with their corporal life and have *common sensibilities*, because without such empathy we have no basis for understanding what we have to feel responsible for and how we should act in order to protect animals. . . . The human responsibility belongs to a temporal and narrative life that cannot be imagined for animals, even the most developed animals.

He envisions that animals live in the present alone, whereas man lives at the same time in the future and the past. This temporality is of a long duration, remembered and anticipated, and includes a capacity to tell a life story that is a product of an all-inclusive imagination. It can conceive a sustainable and viable world that can include the life of future generations and a global context. Indeed, for Kemp, "the responsible human being understands himself as responsible or, rather, as co-responsible for all that he remembers having done and for what he might yet do".

Kemp concludes that

> humans and animals are partners as corporal beings but not as temporal beings; they belong to the same world, but don't have the same duties – one can ask more from a human than from an animal, because humanity can both destroy more and offer more in its relationship with animals than an animal could in its relationship with man.

Hence, humankind has a responsibility to show respect to animals as dignified beings with vulnerabilities and possessing their own (albeit limited when compared with human) sense of integrity and coherent sense of their lives in time and place. Interestingly, as with Taylor, Kemp goes on to claim that this duty of care should extend to plants and all forms of life, although he does not explain this.

It seems to me that this account is instructive in a number of ways. First, it is another argument for respect towards living organisms for their own sake because of their intrinsic integrity and worth. Second, an interesting and important slippage seems to occur throughout the account. Much of its power rests on claims concerning animals being conscious (of their surroundings, purposes, life integrity, etc.) and is at its highest plausibility in the case of the "most developed animals", yet frequently the respect due to animals who possess these attributes seems to get generalized to organisms that do not (for example, arguably, earthworms and plants). Interpreting this seeming slide raises some potentially important issues. For example, is it that, as with Taylor, attributes and capacities normally associated with fairly high levels of consciousness are being conceived as in some meaningful sense attributable to what we would normally consider to be low level or non-conscious organisms? Or, indeed, is there a tacit invitation to revise our understanding of "consciousness" itself, on the grounds that it, too, has become unjustifiably anthropocentric? The repudiation of anthropocentric conceptions of the terminology through which we customarily articulate our worldview – even, as we shall see, the idea of education – appears as a growing theme in some environmental literature, and features in the remaining two views to be reviewed presently. But finally, with regard to Kemp's view, he does make the point that while levels of commonality exist between the lives of the human and the non-human, there are critical differences – and also, perhaps, (and notwithstanding the slippage referred to earlier) that nature is not of a piece: each organism needs to be regarded on its own terms.

I would like now to introduce into the debate two views that seek to extend the attribution of intrinsic value beyond living organisms. This is important because typically environmentalists are not exclusively concerned with the well-being of biological organisms, but also, for example, of landscapes and natural phenomena that have other components. The first of these views – that of Jay McDaniel (1986) – blends theoretical Whiteheadian process philosophy with theological and phenomenological perspectives to argue that *all* existents have intrinsic value. This is a bold view that, if true, would have profound implications for an environmental ethic. It might also be thought to resonate in some respects with the view discussed in the previous chapter of consciousness conceived as the place where (all) things are "let be".

Taking his start from Nobel laureate geneticist Barbara McClintock's "feeling for the organism" as a unique individual, mysterious other, and fellow subject, McDaniel describes how all organisms – which, for him, now includes individual cells and items of non-sentient matter such as stones and molecules – are recognized as ends in themselves in process theology:

> A "creature" can be understood as any entity that is created through its relationships with, and responses to, other entities, including God. A creature may be an energy event within the depths of an atom, an individual molecule, a living cell, or a psychophysical organism such as a deer or a human. In process theology there is no sharp line between creatures that are sentient and those that are insentient. . . . There was never a time in cosmic evolution when all creatures were utterly devoid of sentience, and when, out of sheer inert matter, sentience appeared.

McDaniel claims that a "creature's" responding to external events suggests a reality for itself and therefore "experience". Hence, each creature has its own perspective and intrinsic value. For example:

> The living vibrancy of rocks can be felt in mountains and boulders experienced in certain outdoor experiences . . . in certain forms of sculpture . . . in paintings of artists such as Ruskin and Cezanne.

He cites with approval William Barrett's observation that "Whoever thinks matter is mere inert stuff has not looked long at rocks". And harking back to the discussion of the vibrancy of the native occurring of things in nature developed in Chapter 2, a clear synergy seems to be emerging.

McDaniel goes on to conclude that from a process theology perspective:

> In the last analysis, then, a "feeling for the organism" need not be limited to fellow plants and animals. It can extend to *all* fellow creatures. All existents are, in one way or another, organisms. A feeling for their organismic quality can unfold within the human psyche in many different ways, some of which may seem to the experiencer more "aesthetic" than "religious" given ordinary connotations of the terms. All such feelings will involve a realization that an existent's outer form is an expression of its inward energy and that this energy has a quality of conscious or unconscious for-itself-ness.

He grants that our apprehension of this "for-itself-ness" that is possessed by all existents will often be vague and dim, interfused with other sorts of feelings that obstruct its purity. But, he argues, it remains important as it reveals to us our fellow "creatures" and our solidarity with them, which in turn initiates/requires a new level of respect in our dealings with them. Our world is everywhere populated with existents possessing autonomous purpose and worth. Clearly, ecological justice would require recognition of all of this, again underlining the poverty of mainstream views, but also raising the question of what it would be in such a re-envisioned world for justice to be implemented. Everyday moral judgements become massively complicated if the upshot of this view is that they have to take into account, not "simply" the moral standing of human and near-human subjectivities, but also those of every existent that could be affected.

And yet, perhaps the idea of a value attaching to, or emanating from, sheer existence is not beyond the scope of what is intelligible, and in any case the fact that this might place an infinite ethical demand upon us that we could never fully meet does not invalidate the idea that the demand is present. Indeed, in some ways it only underscores it. Certainly, there are examples of something like value reverberating in our experience of the sheer existence of inanimate things, from a distant constellation to a grain of sand. If we allow ourselves to be so attuned, both can elicit feelings of wonder and a sense of their unique being (also, perhaps of the processes and powers that have shaped them), the absence of which would now be felt as a loss, a diminution of experience. In such moments perhaps here briefly is being foregrounded a dimension of value that lies within us and is fundamental to sensing what is intrinsic, even while it is for the most part occluded from consciousness. It follows that it must be part of a radical ecojustice strategy to excavate what have become dim and deeply buried intuitions that connect us with nature and its value so as to enable them to shine through as part of the occurring of things. Such enrichment of experience would be central to the falling away of overly instrumental consciousness and the metaphysics of mastery. And, at the very least, a case is made both for the need for a significant modification of current everyday consciousness if a proper attunement with nature is to be achieved, and for the importance of moments of heightened awareness in energizing this.

The final view that I consider resonates with some of the aforementioned, but takes as its central theme the development of a non-anthropocentric idea of learning and its implications for how we should relate to the natural world.

Ramsey Affifi (2017) argues that there is a "metabolic core" to environmental education. He works with an "enactivist" account of cognition that sees it as the process of an active embodied organism rather than the mental activity of a deliberative consciousness. Here distinction-making is not based upon a representation of some external reality, rather it is "constituted by the needs and organization of the organism's embodied adjustments to its world" – somewhat as implied in the aforementioned views of Taylor and Rolston. For example, being a nutrient is not inherent in the physicochemical nature of the world: it is a relational feature. It depends on another organism utilizing it as a nutrient. He argues:

> Because distinction-making is driven by an organism's history and needs, distinctions have a significance, and enact a semiotic domain of goals, valence, meaning, threats, possibilities and other qualities in the world. . . . Metabolism *does* distinction-making, first by enacting the distinction between organism and world, and thereafter by enacting distinctions within each that enable the organism to continue to flourish. . . . *[L]earning is the process of changing distinctions* that is both the condition and effect of the metabolic organization of all life forms down to, and including, single cells.

In this overall context, space and time become horizons with the organism at their centre and they extend outward, graded by the organism's needs and concerns. A striving to meet goals through space and time "places a teleological matrix overtop the physicochemical world". He gives the example of an *Escherichia coli* bacteria swimming up a sugar gradient towards a food source. The organism recognizes and responds to *differences* by comparing states in space and time and acting on the basis of what these comparisons signify for its purposes. Here perception and action are not separate, rather this is all one process. Information is *enacted*. Affifi avers that this self-constituting organization of the living organism gives it an ontological status that is not reducible to its underlying components and that this constitutes the "true meaning of metabolism".

This argument has interesting implications. For example, learning as defined earlier means that through organismic metabolic activity, organisms disclose an environment that *educates* them – and they educate their environment (that contains other organisms). Primordially, the environment, then, is an educational organization – all organisms exist in mutually educative relationships. For Affifi, this is the fundamental meaning of environmental education. And it requires humans to participate openly in this dynamic interplay and to uncover the educational organization of ecosystems. Affifi exhorts us to learn how to silence our chattering and to listen to the non-human world, and also to redraw how we talk about "biology", "education", "culture", and related concepts such that human supremacist conceptions of cognition, learning, education, etc. are replaced by enactive ecological conceptions as the touchstone of thinking. When this happens, the continuity between humans and other organisms is recognized, and, therewith, their inherent intrinsic value.

Here, then, we are provided with a rich vision of nature's integrity as a milieu of intricate, complex, evolving "learning" relationships in which all life is engaged and human life is embedded. There is surely a sense in which this vision intimates an important aspect of the natural world. But on Affifi's account, to enter it, we are required to consider a radical reworking of concepts that are fundamental to much mainstream thinking – indeed, it seems that a pretty radical reconfiguration of our current form of sensibility is a precondition of achieving a proper attunement to nature. While undoubtedly there are good arguments expressing a need for deep cultural change that, *inter alia*, facilitates a passivity on our part that allows us to hear nature's voice and thus be "educated" by it, it is not clear that a fundamental and wholesale renunciation (or at least relegation to some secondary status) of our existing form of sensibility is either necessary or helpful to its achievement. This is partly because change of this magnitude is deeply problematic (as will be illustrated later) and partly because an elevation to a central orientating position in our interpretation of the world of purely "enactive" versions of cognition, feeling, agency, value, etc. surely is deeply reductive. Shorn of the rich resonances of meaning available to intentional consciousness, these terms become severely attenuated – indeed, they begin to appear as mere outriders or proxies for a world now primarily conceived in terms of causality and stimulus–response theory. It seems to me that the banishment of what we currently understand by these terms from a central position in making sense of the world is a form of scientism that, while encouraging a new attentiveness to nature, threatens to diminish our ways of relating to it – and also our understanding of ourselves as participants in this relationship.

Notwithstanding that in places I have offered some serious reservations, I suggest that in the context of elucidating nature's intrinsic value, and taken as a whole, the preceding accounts express some significant insights and raise some important considerations and issues that can be summarized as follows.

First, they all radically question and seek to reveal what they take to be the harmful consequences of a pervasive anthropocentrism in our perception of the natural world. They argue that anthropocentrism distorts our perception of nature, cabining it so that things in nature are only very partially revealed to us, many of their most important facets remaining invisible. This leads to actions that are fundamentally ignorant and arrogant. In order to overcome this, we need to develop an attentiveness that is attuned to what lies before us rather than what we project upon it. This has radical implications. Its achievement requires that in various ways, we jettison both our customary anthropocentric classifications of the things that we encounter and be prepared to re-evaluate our interpretations of many ideas that are primal in our conceptions of our relationship with the world, such as "cognition", "value", "identity", "experience", "sentience", forms of existence, and, indeed, "education". Undoubtedly, there is much that deserves attention here.

Second, viewed from a conventional standpoint, central to much of this demand for radical reconceptualization is a redistribution of properties and capacities that we normally attribute to consciousness: a generalizing of capacities such as perception and cognition that we normally understand in association with sentient creatures (and according to what we take to be their level of consciousness) to non-sentient existents such as plants, or on some views, stones and atoms. Or, to put it more accurately, we must either stop conceiving these capacities exclusively from the perspective of consciousness altogether, or radically change our understanding of the range of existents that can be regarded as possessing consciousness in some meaningful sense.[4] That is to say, that we should cease to regard human consciousness as setting the benchmark for what counts as consciousness and therewith what is the true meaning of terms such as "value", "agency", and "cognition". Either way, this raises the critical issue of how we are to understand these key terms when they have become reconceptualized in this way. Here I think that a number of serious problems arise. For example, to what extent can meanings and significances that comprise or are associated with a term in one context legitimately carry over to another? It is one thing to agree that we should not hastily dismiss ideas that run counter to our cognitive habits, but we still have to *understand* them, make sense of them. And while this is likely to require some reorientation of our current cognitive resources, it cannot be done completely independent of them. This is especially salient when the reorientation demanded is at a level so fundamental as to effectively dismantle our underlying form of sensibility, as indicated earlier. Some sort of responsible judgement needs to be made concerning the intelligibility of new ideas in relation to these pre-existing interpretive structures and this will include exploring where innovations are problematic and what the cognitive (and other) cost of admitting them to our worldview might be.

To take one example of the difficulties that can arise: in his ethics of respect for nature, Taylor argues that a key rule is not to attempt to deceive animals that are capable of being deceived by moral agents, and that the clearest and commonest transgressions of this rule occur in hunting, trapping, and fishing. Taking the latter case, we might ask: what level of consciousness is being attributed to fish? Is not a full idea of deceit bound up with mainstream conceptions of consciousness where anticipation and expectation are *experienced* and can be experienced as intentionally induced and confounded by a deceiver? Morally culpable deceit occurs in the realm of intentionality, a realm shared by deceiver and deceived. This can only occur for what Kemp describes as full temporal beings. On this account, deceit cannot occur within nature and nature cannot deceive us. We can be *mistaken* about some aspect of nature, make erroneous judgments about it, but not be truly deceived *by* it. For it to be otherwise a questionable anthropomorphism would appear to be in play. And, more generally, is it not the case that along with deceit the meanings of other terms taken to have application in nature such as cognition and education (Affifi) are rooted in the milieu of human consciousness? To reapply them to non-human contexts is problematic and perhaps involves a kind of arrogance – precisely of the colonizing kind that advocates of this position deplore. To return to the aforementioned example, *if* fish think, we do not know what they think and should not presume to do so. And if, say, we attribute a purely enactivist kind of thinking to them, as advocated by Affifi, morally culpable deceit does not arise. A fish taking a bated hook is no more deceived by us than is a plant subjected to artificial lighting to encourage flowering. This is not to say that we should not be circumspect about the possibility of causing pain or of interfering with natural processes, but rather that such (welcome) sensitivity is hardly enhanced when elucidated in a terminology that remains emotionally freighted while becoming – as I have argued – cognitively reductive or empty.

Yet this impulse to find commonality with aspects of the natural world has a worthy origin that needs to be recognized: much human behaviour towards nature does not exhibit the respect that is its due. Undoubtedly, this can be the result of separating humankind from nature and turning the latter into a homogeneous unknowable, yet inferior, amalgam that is set up to be colonized – a point made by Affifi and others in a paper to be discussed in Chapter 7. But, perhaps, the respect due to nature is not because of what it shares with us, but because of its own integrity, otherness. In Chapter 2, I argued that there is purpose and agency in nature, but its precise character is inherently mysterious and is effaced when attempts are made to reduce it to something that it is held to have in common with the human. There is a level at which the overriding need is to accept the mystery in things in their arising, rather than to dissolve it through attempts at conceptual re-engineering.

Because it is central to our current theme, I give one more example of the difficulties that arise when the attempt is made to transpose fundamental concepts in the ways described previously: that of "value", or more accurately, "valuing". When Taylor and Rolston III speak of, say, a plant valuing its life, how does the meaning of valuing here compare with when we speak of a person valuing their life? Again, is not a set of significances warranted in one context being surreptitiously imported into the other? Rolston III argues that in maintaining themselves, plants value in the only way available to them: "botanically". Now surely this is to acknowledge that such valuing is of a very different kind to human valuing and that therefore through extending the use of the term in this way, the logical implications of human valuing are forfeited. Given this, what implications remain for the kind of respect that is due to plants? If the touchstone of human subjectivity is removed from the idea of valuing, are we not left with a mere chimera? In which case, once any such illegitimate transfer is detected and eschewed, those wishing to use the argument to enhance the regard in which, say, a plant is held, will fail in their purpose.

It seems to me that with regard to warranting appropriate respect towards nature, it is better to leave our fundamental ideas of consciousness alone, to stop relying on arguments that purport either to demonstrate a continuum of consciousness between humankind and the rest of nature, or to eschew all peculiarly human resonances from either the idea of consciousness or concepts that are taken to characterize it. When pressed, these arguments are in severe danger of becoming inordinately complicated and ultimately futile. Rather, let us focus on the otherness of nature, and what phenomenology can reveal. Perhaps important concerns that motivate the views that I have been discussing can be retrieved in the course of this approach. Before pursuing this, I wish to make two preliminary observations on the idea of nature possessing inherent intrinsic value that are prompted by the preceding discussion.

First, with regard to nature, phenomenologically the kind of intrinsic value that it emits varies; it is not monolithic. We do not have to choose between different kinds of intrinsic value and assert just one as pivotal. This is because nature itself is multiform and dynamic such that our experiences of it may present different aspects that are equally "true" depending on context. In some cases or respects, nature's intrinsic value might, in conventional terms, be regarded as purely moral, in others, as purely aesthetic. For example, whereas the intrinsic value of a higher ape might derive primarily from the quality and integrity of the conscious life that it is taken to enjoy, the value of a landscape might be experienced through its integrity as a field of ambience: it gives off significances that participate in one another in some special way, arising out of a particular set of mutually sustaining relationships of the kind attributed to "places" in an earlier chapter. It might be philosophically tidy and satisfying to identify one fundamental source of value that underlies all others, but in the case of nature, this ambition has to be relinquished. Because of nature's manifold presence, it is both inappropriate and itself expresses precisely the drive for (now intellectual) mastery that it has been argued frequently bedevils our engagement with nature in the West.

Second, within a phenomenological approach, the existence and character of any moral standing and intrinsic value that nature might possess are disclosed in our direct experiences of nature. Because these are ineluctably *human* experiences, attention to them can reveal how a proper respect for nature involves recognizing both its alterity and its relationship to humanity. Here a path is opened towards a non-anthropocentric, yet deeply human, understanding of nature and its value. Amongst other things, this might reduce the antagonism that some see to be inherent in the quite legitimate questioning of the pervasive assumption that humankind should be valued above non-humankind – provoked, it must be remembered, by a situation in which the activities of the former are resulting in mass extinctions of the latter.

Nature, place, and normativity

In their different ways, the views that I have been considering can be interpreted as drawing attention to some important intuitions regarding nature: that things in nature can be sensed as having an integrity and good of their own; that purpose and agency is present in the natural world; that being able to perceive this can involve a heightening of consciousness and can require a loosening of the grip of some of our conceptual structures and instrumental purposes; that there is a possibility of participating in the natural world in more receptive ways and of listening to what we can think of as nature's "voice" in its otherness; that sheer existence – the original standing forth of things – can be felt to emanate significance and value such that even non-sentient matter is not merely inert and can be the source of experience rich in meaning. Taken together, these intuitions express an important reorientation in experience that is necessary to a fuller – that is to say, truly emplaced – understanding of our environmental situation. They also resonate strongly with the view that I have developed in earlier chapters and help to foreground some important emphases that it might be helpful to summarize.

In Chapter 4, I explored the ecstatic nature of consciousness. There it was argued that it is our ability to live beyond ourselves, *with* the things of which we are conscious, that enables both the reflexivity necessary for self-awareness and the sensing of our emplacement in a world – a world of infinitely extended and interconnected significances made manifest to us through the emplaced occurring of things. Here things in nature are sensed as proceeding on their own ineffable way, and whose sheer existence is essentially mysterious. It is as little possible for thought fully to encompass the being of a harebell silently emerging through the grass underfoot as it is for it fully to encompass the being of the Orion Nebula. Such being with things and the myriad demands that it places upon us is what it means to be in the world. Conceived in this way, places are imbued with feeling and value; they are inherently normative in the sense of offering intimations of possibilities of right and wrong conduct. These are expressed in the ways of proceeding that they invite, the attentions that they demand, the responsibilities that they offer, and the mysteries that they intimate.

In Chapter 2, I explored the particular character of the occurring of things in nature, deriving from their self-arising quality and an ontology of mutually sustaining relationships in which their mystery and address are heightened in experience. Here the normative intimations and sense of intrinsic worth are experienced as emanating from their own transcendent particularity and otherness. Through our embodied participation in the being of places exemplified in that chapter by the upland stream and the woodland dell – by entering into their otherness – our being is enriched and quickened. Our ecstatic consciousness, as inherently environmental, is fulfilled and we are able to appraise the world and discern appropriate action from a perspective that in a significant sense lies within nature, one that is enlightened by its internal norms. Here is an important sense in which we retrieve our place in nature – as it were, become part of it as its conscious expression. It seems to me that this offers an important interpretation of what lies at the heart of what it is to be *in* nature – an issue raised in Chapter 1 and taken to be pivotal to the idea of ecologizing education.

Intrinsic value in nature and the principle of social justice

If the preceding account is taken seriously, then it follows that when it comes to the issue of justice, ingrained anthropocentrism has severely blinkered our understanding of its full possibilities and requirements. If it can be shown that nature – or at least significant aspects of it – possess an *inherent* integrity and intrinsic value (that is to say, intrinsic value that is not simply the product of human beings conferring it) and that, properly attuned, we can become aware of these qualities, there would seem to be extensive implications for morality in general and the principle of social justice in particular. Inherent intrinsic value enjoins respect and if current versions of justice focus exclusively on respect for human beings, they require radical modification. The traditional ethical position of conceiving moral obligation as something that holds only between rational agents can no longer be maintained and the prominence given to social justice in our attempts to order society is no longer credible – partly now because an extended understanding of the community of which we are a part has been retrieved.

In the light of this, the idea of ecological justice requires us to reappraise the basis for thinking about equality and fairness in the distribution of global resources. It also marks the passing of an era in that whereas for millennia it has been nature that has largely determined this distribution, now, in many respects, it is humankind. The nature needs half movement based its position on scientific estimates that in most regions and, on average, around 50% of the environment would require strict protection from human activity in order to maintain full biodiversity (Kopnina, 2016). While the estimates involved here might be considered to be very generalized and result in what might look like a rather arbitrary, but convenient, figure for a slogan, a highly salient point is at stake: the claims of nature have to be taken seriously and there is something deeply wrong morally about sacrificing to the point of extinction many other species in the interests of satisfying the ever-growing wants and unremitting spread of one. Anthropocentrism or human supremacism taken to this degree goes against both some basic intuitions concerning the worth, beauty, integrity, and nobility of what is being decimated and also, arguably, is self-defeating once quality of life is asserted against simply quantity.[5] This is because, if previous argument is correct, participating in such decimation contradicts our own human essence as previously characterized.

Striking a balance between human well-being and the well-being of the natural world involves the exploration of a complex and tensioned relationship. There are important senses in which each requires the other. While there surely is a sense in which nature thrived before the arrival of humanoids, there is also an important sense in which it did not. Before the evolution of human consciousness (or its near equivalent[6]) – and notwithstanding some of the arguments discussed previously – there was no place for nature to be revealed in its many facets, for it to achieve fuller *significance*. Only with temporal beings can it become part of a narrative in which its worth, beauty, integrity, and nobility are capable of being recognized, articulated, and celebrated. In this regard, human consciousness and nature arise in a dance of mutual sustaining. This requires both on the part of humankind a letting be of nature and on the part of nature an accommodation of human needs (as distinct from mere wants) to utilize nature and to defend itself against disease and pestilence. All of this to occur within a realm of responsibility and sensitivity to the moral and aesthetic intimations that emplaced consciousness can receive – even as we may suppose that, for the most part,[7] nature is perfectly indifferent to human well-being.

Previously, I argued that human consciousness is at some deep level orientated towards natural things and the powers that ground and quicken them, these providing an ultimate reference for sense-making. But even if this argument is rejected, insofar as direct experience of nature is an important contributor to the enjoyment and well-being of at least a significant number of people (see, for example, Louv, 2010) unfettered population growth (both local and global) is a serious threat because of the natural habitat destruction that results. The details of this claim are highly complex and in many respects contestable. Indeed, in their discussions of human well-being, many make no mention of population growth as a relevant consideration. Yet, in a situation of burgeoning growth and (at any one time) limited resources, human population growth and human well-being cannot be decoupled. There is an important sense in which all environmental degradation is a product of human population growth: pollution, habitat destruction, water shortage, and so forth are not simply a product of size of "footprint" but of number of "feet".

For example – and harking back to the perspective of the principle of social justice – *if* the ecological carrying capacity of the planet is being exceeded, redistribution of resources to attain equality (even if politically it were achievable in practice) would be no answer to our environmental predicament, given that, as seems likely in the light of the material aspirations of emerging economies and prevalent ideas of "levelling up", overall consumption will not be reduced. For present purposes, I simply note this as an important concern. Resolutions of it potentially have extensive educational implications, one of the lesser of which (but still very important) would be the incorporation of parenting and family planning into the curriculum and in some cases the education and empowerment of women in the appropriation of their bodies, and discouragement of irresponsible procreation. In a democratic society, a greater part would be an incorporation of serious exploration and discussion of human well-being in an age in which natural systems are under severe threat and their moral standing is insufficiently recognized. Some of these concerns overlap with the concerns of those who fight for social justice, but the exclusive focus on human society and the way in which this nuances the issues has to be rejected.

One of the effects of a burgeoning human population and the megalopolitan conurbations that arise to accommodate it is the increasing ontological distance that opens up between human beings so situated and nature. This includes the loss of untimed direct emplaced acquaintanceship with things in the natural world. Here experiences that are held to be pivotal in establishing a right relationship with nature become more attenuated. It is not simply the reduction and effacement of natural phenomena in such circumstances that is the problem, but the focus of life – its distracting busy-ness and the tacit inurement to and frequent glorification of the artefactual, along with a disdain for the (naturally) given, as referenced in Chapter 3. Here, we are returned to the metaphysics of mastery and our attention is called to the limits and dangers of too exclusive an alignment with the principle of social justice. As it gains momentum – and particularly when, as frequently is the case, it is calibrated largely in terms of materialist outcomes – social justice exhorts us to take everything into our own hands – order all – so that everything is available for equal distribution according to (someone's version of) what is rational. Ultimately this requires that somehow everything of value be quantified and hence fixed, objectified, and categorized, its "being" modulated in a way that makes it disposable and accessible to all in accordance with some rational plan. Here, consistent with the metaphysics of mastery that it expresses, an irony occurs. This process inevitably subverts the individuality and particularity of the existents that are to be redistributed or made available, wrenching them from their strange emplaced existence and hence destroying genuine diversity. Ontologically reconstituted in this way, the qualities of engagement that they can offer become severely impoverished.

The idea of ecological justice and its tensions with the anthropocentrism expressed in mainstream notions of social justice raise fundamental issues concerning the nature of morality. How these issues are resolved has extensive significance for both our understanding of moral/social education and education itself. Along with others, these matters will be pursued in the chapters that follow. For the moment, I make the point that the position at which we have now arrived returns to centre stage a number of previously raised key questions: what really should count as human needs? What demands can we *properly* make upon nature – that is to say what demands are we warranted in making in the light of an acknowledgement of nature's own needs? Presumably, there will not be one final objective answer to this question, as, *inter alia*, such legitimate demands will always be relative to, say, the current state of nature and in some respects will be determined by it. For example, presumably a legitimate demand cannot outstrip what nature can (sustainably?) provide. Furthermore, there are likely to be significantly different cultural perspectives on the issues that are raised and in many ways it would go against the grain of the current thesis to attempt to homogenize these perspectives – indeed, the danger of a form of eco-fascism lies here.

Nonetheless, it remains important to pose these questions, for simply to ask them is radically to reorientate the current environmental debate and places renewed emphasis on the importance of how we are to receive and understand nature's requirements. The character of such knowledge and the issue of nature's having a voice to which we can become attuned are explored in the next chapter. But for the moment, I wish to emphasize the need to make a start on addressing these questions. It will be interesting and important to identify where there is a significant degree of consensus and where there are significant differences that require further discussion, and the sorts of compromises that might be possible at any one point in time. By its very nature, even if a broad and rather abstract articulation of what humanity can rightly require of nature is achieved (and certainly some broad overall global sense of direction is of the essence), at the level of practical detail in varying cultural, geographic, and economic contexts, of necessity this will be an ongoing debate. Before proceeding to a deeper examination of the underlying terms of this enquiry as conditioned by an enhanced understanding of the character of a proper knowledge of nature, it will be helpful to consolidate the argument as it now stands – particularly regarding its position on the issue of anthropocentrism.

Anthropocentrism reconsidered

One of the key features of the argument developed in this chapter has been an ongoing critique of anthropocentric approaches to nature and environmental issues. Yet there are many who would question what might appear to be a pretty wholesale repudiation of anthropocentrism, find it puzzling – even morally questionable in that in some contexts it might appear "anthropophobic" and to challenge ideas of the sanctity of human life. Furthermore, does not the idea of environmental consciousness championed in this book involve ineluctable anthropocentric elements in its developed account of intentionality in that it is within the space of human consciousness that the occurring of things in nature is envisaged? And is it plausible to posit an approach to environmental issues that seems to relegate the interests of our own species by placing them on a par with those of any other species? These are very relevant questions and responding to them will help both to defend and to refine the position that I am espousing. I will begin by gathering some key strands of a defence of anthropocentrism as a response to environmental issues.

One potentially powerful rejoinder to my critique of anthropocentrism in this context might be that in many respects we are unavoidably anthropocentric in our outlook: we cannot but help see the world from a human perspective because, amongst other things, our human form of sensibility is a key determinant of what we perceive and as far as we know this form of sensibility is unique to human beings. Indeed, has not the thesis that I have been developing laid stress on the otherness of nature? Undeniably, there is a good deal of truth in this response. For example, we know that our senses do not correspond exactly with those of other sentient beings and that they filter as well as reveal. In addition, we believe that the human intellect is capable both of drawing distinctions and of levels and kinds of reasoning that exceed that to be found anywhere else in nature. These are significantly the products of ever-evolving human culture(s), and when they are allied with the reflectivity and temporal nature of human consciousness (as earlier referenced in the discussion of Peter Kemp's views) they completely transform the character of our experience compared with that of the rest of nature.

To give just one example, we do not conceive of, say, chimpanzees as being capable of sitting down and thinking about what they will do the day after tomorrow, nor to be capable of articulating and intentionally choosing to follow a moral code. This is not to say that there is a completely unbridgeable gulf between all aspects of human experience and that of other sentient beings. For example, the behaviour of all sentient beings suggests that they share a world of solid objects that possesses at least some features that are common to us. Indeed, some possess sensory apparatus that is far superior to ours in particular respects and certainly some appear to exhibit a degree of reasoning that is recognizable to us. But as far as we know for the aforementioned reasons – and others – human consciousness is significantly different from all other forms, and therefore our perspective is unique and cannot be transcended by us. We are epistemologically and ontologically unique, and while there is clearly a sense in which equally this could be said of every different form of sentient life, the point is surely made that here there is an important sense in which we cannot "escape" our anthropocentrism.

If, once stated, this point seems blindingly obvious, its implications perhaps are not. For example, it does not follow from any of the previous discussion that, as with modernist humanism, humanity necessarily should be regarded in some general way as a superior kind of being set apart from and above the rest of nature, and that it is the only kind of being capable of possessing intrinsic value. Nor does it support the view that even if it is allowed that other beings in nature have intrinsic value, the desires of human beings always trump this and that therefore anthropocentrism becomes not merely epistemological and ontological but also ethical – that is to say, anthropocentrism is a justified ethical position to take with regard to the environment. If it is granted that nature (or some aspects of it) has intrinsic value – and therefore has moral standing – then unqualified anthropocentrism as an ethical position can never be adequate.

But perhaps the defence of anthropocentrism is a little less "high flown" and principled as this. Maybe the argument is that, just as with other species, it is "unnatural" to put the welfare of other species above that of one's own. After all, on a Darwinian view, does not each species strive for its own survival in competition with others? In biological terms, its overriding purpose is its own reproductive success, and as Darwin puts it in *The Origin of Species* (1872 edition):

> If it could be proven that any part of the structure of any one species had been formed for the exclusive good of another species it would annihilate my theory, for such could not have been produced by natural selection.
>
> *(p. 196)*

This being the case, is it not best simply to accept as hard biological "fact" that ultimately we will place our own interests above those of other species and, given this, rely on enlightened human self-interest to save the planet? In which case the best strategy for those concerned about the welfare of nature is to show how it aligns with human welfare. Undoubtedly, this is (and always has been) a powerful voice in much environmental debate, and the way in which the recent emergence of likely disastrous consequences of global warming for human flourishing seems to be spurring political action is a cogent illustration of this. Certainly, tactically those who care for nature would not wish to discourage this voice. There is an obvious truth underlying it. But to have this voice completely dominate debate would itself be a disaster. It would play into the underlying orientation given by the metaphysics of mastery with all its distortions of our fundamental relationship with nature and the ultimately dire consequences that follow from this.

The arguments in this area can become very complex and, indeed, convoluted. Perhaps much of this can be circumvented if we return to the issue of what counts as human flourishing and the suggestion that a celebratory relationship with nature is a key component. If, indeed, human consciousness is ecstatic in the sense elucidated in the previous chapter, and if the argument set out there is accepted that there are important and irreplaceable ways in which it finds fulfilment through a poetic engagement with things in nature in their native occurring, then there arises an important sense in which the flourishing of human consciousness is not merely contingently but ontologically aligned with that of the natural world.

But what of those who, drawing on their personal experience, reject this way of looking at things and who, for example, find great satisfaction in the realm of the artefactual, and perhaps would see it as a matter of only passing regret that significant denaturalization of the planet went on apace if that were to be the price of continuing or expanding this way of life? One might imagine someone whose life centred on perhaps socializing, electric urban blues music, some academic discipline, or chess, who would happily adopt this attitude, and whose "enlightened self-interest" would consist in protecting the natural environment only to the degree necessary to preserve the viability of their participating in their chosen activities. What, if anything, would be wrong with this? From the point of view of a lifestyle choice within a liberal democracy in which, within the framework of liberal values, each individual should be free to choose their own version of the good life, perhaps nothing. It would appear to be both perverse (at least in the short term) and authoritarian to condemn those who find complete fulfilment exclusively in the company of others and perhaps the energy of high rise "big city" life. But from a strictly ethical point of view – and from the perspective that I have argued – this position is revealed as impoverished: it fails to take account of all entities that have moral standing and has become insensitive to the fulfilment to be had in engaging openly and respectfully with them. If, as previously argued, nature possesses inherent intrinsic value – that is autonomous value that arises from within itself and is an integral part of its native occurring, as against being conferred by human beings – this value demands respectful consideration by moral beings. It should be noted that, as set out here, this argument stands even if it should turn out to be the case that city dwellers produce a smaller environmental "footprint" than rural dwellers. Sustainability, as characterized in this book, is not essentially about the level of material consumption of nature, but the quality of an individual's engagement with it – although, as has been made plain, this relationship itself is envisaged as essentially non-consumptive in the sense of not regarding the natural world primarily as a commodity.

Notes

1 Ecological carrying capacity is defined as the maximum population size of the species that the environment can sustain indefinitely. Amongst other things, clearly this involves estimates of levels of consumption and pollution incurred, and, in the case of the human species, what would count as the material and social needs to service an acceptable quality of life. In other words, in the case of human beings, ecological carrying capacity becomes modulated in terms of an idea of an *optimum* population that can be sustained indefinitely. To be sure, this latter will be a matter of considerable debate and this is fully acknowledged by Daily *et al.* who set out what they consider to be the key considerations concerning an acceptable quality of life in their paper. They also point out that even if they have underestimated this optimum by 100% and it is actually 4 billion, current and projected populations remain well beyond what can be sustained indefinitely and will have deleterious effects on the flourishing of future generations.

Of course, none of this should be read as minimizing the range of very sensitive moral/religious/cultural issues that the idea of population control raises in many contexts. Nor does it deny the difficult economic and social issues that, for example, an inverted age structure that results from a fertility rate of less than 2.1 would produce. The argument is that the issues raised are indeed difficult and complex *and that they need to be addressed.*

2 There are, of course, varying notions of social justice, but for present purposes the salient point is that they all share an exclusive preoccupation with *human* well-being.

3 I include the term "inherent" to distinguish this from an anthropocentric version of intrinsic value that refers to the value attributed to things that humans find satisfying *per se* rather than for instrumental reasons. Although in a sense "intrinsic", fundamentally its existence is entirely dependent on the tastes of a valuer (see Bonnett, 2004, pp. 78–80).

4 There are resonances here with many indigenous views of reality – views that have arisen during the course of prolonged and intimate contact with nature.

5 It is important to note that at this point, the term "quality of life" is intended in the broadest sense of life enrichment that, in line with previous argument, should not be equated simply with current Western conceptions and would embrace, for example, indigenous perspectives.
6 I use this term in order to allow for the distinct possibility that other creatures such as other Simians and some Cetaceans, might be capable of significant degrees of self-awareness, reflective choice, and empathy.
7 "Most part" because this leaves it open as to whether some "higher" animals and domesticated animals possess this ability to care to some degree.

6

LISTENING TO NATURE

Ecological truth and systemic wisdom

In the previous chapter, the sense in which nature possesses intrinsic worth and hence moral standing was reasserted in a way that preserved its distinctiveness from human consciousness, and that found expression in a notion of ecological justice that displaces highly anthropocentric notions of social justice with the idea that nature, too, can be understood as having needs that demand to be taken into account. In the light of the characteristics salient in knowing nature explored in earlier chapters, this chapter discusses some key elements involved in listening to nature and how these raise important considerations for how we conceive knowledge, truth, and wisdom in education more generally. The idea of becoming attuned to nature's "voice" is explored and an ontological notion of truth is developed in which the focus is not exclusively upon the objects of knowledge, but also upon how we *are* towards the things that we know. These ideas and the considerations to which they lead represent a further elaboration of what might be meant by environmental consciousness and become foundational to the ideas of ecologizing education that are explored in the final chapter.

Nature's voice

A key issue that arises from the preceding account is the sense(s), if any, in which it is proper to conceive of nature as somehow expressing its inherent integrity, intrinsic value, and normativity. That is to say, in what senses is it proper to consider nature as having a voice. From a moral perspective, it is hard to overestimate the significance of this consideration, for if it is possible for us to "hear" this voice – to discern the integrity, intrinsic value, and normativity that nature emits – this would need to be recognized in all our dealings with nature. It would constitute a fundamental element of our relationship with nature and would require from us an ongoing – and perhaps, frequently ethical – response. In assessing the terms in which environmental issues currently are debated, Pall Skulason (2015) argues that we need to nourish a spiritual understanding of nature by developing "our capacity to receive and understand the messages that nature is sending us" such that "we are able to form our life in accordance with the meaning of Nature herself". For some, this line of thought is highly problematic and, indeed, fraught with danger. As we have seen, the natural sciences portray nature as ultimately a value-free realm of fundamental particles and forces in which the idea of such a voice would be nonsense. Does not their success in revealing and predicting the physical world demand that their portrayal of nature should not be easily overridden?

Certainly, some prominent environmentalists have felt this. For example, and with regard to the idea of nature possessing intrinsic value, in order to remain consistent with the assumption made by "scientific naturalism" of an objective value-free natural world, J. Baird Callicot (1986) argued for what he called a "truncated" notion of intrinsic value. This argument held that while the *source* of all value is human consciousness, it by no means follows that the *locus* of all value is consciousness itself, or some mode of consciousness such as reason, pleasure, or knowledge. We are perfectly capable of valuing – and typically do so value – things other than ourselves, including aspects of the natural world. Furthermore, claiming that something acquires value because someone values it does not prevent it being valued for itself (rather than, say, being valued for instrumental reasons). It simply means that an intrinsically valuable thing is valuable *for* itself, but is not valuable *in* itself. Here, then, we have an anthropogenic notion of intrinsic value that would be consistent with scientific naturalism. Nature can have intrinsic value, but this value is not inherent in nature itself, rather, it is conferred by human subjectivity. In an attempt to avoid the "emotive relativism" that the reference to human subjectivity might seem to entail, Callicott appeals to Hume's idea of a "consensus of feeling" – i.e. the idea that the human psychological profile is standardized in crucial respects. Hence, other things being equal, when properly sensitized, in broad terms, humanity will ascribe similar intrinsic value to the things that it encounters. In this way, a kind of objectivity is achieved.

But now a question arises as to why there should be this generalized acceptance of a need to accede to the requirements of scientific naturalism. At its philosophical inception, Lyotard (1984) was famous for noting postmodernism's incredulity over the acceptance of grand narratives. In this regard, while it might be entirely proper for science to explore what can be revealed when the world is construed as purely physical, does not Callicott's acceptance of this construal when we move into the ethical sphere represent just such an obedience to a grand narrative that, if not questionable within the sphere of science, certainly is so when applied outside this sphere? Surely, here scientism raises its head again, for the view of nature that is privileged by scientific naturalism is an integral part of that constellation of ideas described in Chapter 1 and that I dubbed "modernist humanism". This constellation functions as a grand narrative of considerable ambition and hubris – and its view of nature marginalizes widespread culturally recognized experiences of the natural world in which value can be felt as inherent in, and to radiate from, the things encountered. As such, surely the generalization of scientific naturalism should be subject to a degree of incredulity. Indeed, previous argument indicates that the narratives of modernist humanism deeply misrepresent both nature and the essence of experiencing value. Regarding the former, examples of our experience of nature have been offered to show that nature *itself* is far from being a value-free realm. Regarding the latter, it has been shown that essential to nature's intrinsic value is that it is inherent: phenomenologically we do not simply confer/project this value, but discern it, receive it. Often, simply, we are *struck* by it, as perhaps when, say, we glance up at the sky on a clear night. This makes a fundamental difference to our stance towards, and perception of, nature. Rather than one of masterful projection, it is one that instantiates a degree of submissive acceptance. We are both enlivened and humbled by the beauty and grandeur that is conveyed to us in this perceptual experience. And, in perhaps less elevated experiences of nature, we are sensitive to what it has to tell us through encounters with its everyday small-scale occurrences such as a bud's breaking open or a bird's squarking. The further significance of this sensitivity to what I have termed *inherent* intrinsic value as contrasted with *ascribed* intrinsic value will be explored presently.

But suppose that, for a moment, we set aside this issue concerning the idea of a value-laden nature not being consistent with scientific assumptions: are there not in any case large dangers in ascribing any moral authority to nature? For example, whose interpretation of nature and whose interpretation of its "voice" is to be followed? There is a long history of what is, and what is not, considered to be natural, or in keeping with the natural order, being used to authenticate practices that, for example, maintain social hierarchies, criminalize "deviant" behaviours, and discriminate in ways that have later become regarded as arbitrary or spurious. Clearly, there is a stringent need to remain alert to possibilities of what is taken to be nature's voice being recruited in ways of this kind – and not only in the human–human context, but also in the human–nature context. Here, for example, what some have "heard" (such as the infinite superiority of human beings and their "needs" over the rest of nature) can be used to justify behaviour that inflicts extreme pain and hardship on fellow Earth-dwellers, for example, in the sphere of medical research, and indeed at one time, cosmetics research. This is not to mention the wholesale decimation of habitats and environments in the pursuit of an inexhaustible demand for energy and materials. Here nature is "heard" as only a resource. How are we to identify nature's true voice from amongst what might be a plethora of imposters?

In part, considerations of this kind take us back to views concerning the ways in which we can experience communication with nature developed in previous chapters. There, emphasis was laid upon the way in which through participation in the native occurring of things we can become aware of nature's integrity, agency, intrinsic value, and normativity. These supply a sense of what would or would not be a fitting response. A particularly well-argued exploration of the philosophical grounds upon which nature can be said to have a voice is provided in James Magrini's book *Ethical Responses to Nature's Call: Reticent Imperatives* (2019). This carefully argued account resonates strongly with the views developed in earlier chapters of this book, and, indeed, in some respects can be seen as developing these views further in important ways. Hence, it will be helpful to recount some of its main themes and arguments.

Taking an eco-phenomenological approach, Magrini develops the argument that nature radiates imperatives to which we need to become attuned. However, these imperatives are reticent in the sense that their normative message lies hidden beneath the surface noise of things. For example, it is effaced by the naturalistic attitude described previously, and that in turn he sees as part of the current historical epoch or *Weltanschauung* of what he terms "Secular Humanism". (I take this latter to be closely allied to what I have dubbed "modernist humanism".) Drawing on the thinking of prominent phenomenologists and ethicists, Magrini seeks to show that proper attention to things in nature and the ontological themes that arise from elucidating their character can provoke a re-attuned understanding of nature that lies outside of, or is antecedent to, the grip of Secular Humanism. When this occurs, things in nature are revealed as possessing qualities that intrigue and fascinate, draw us to them through their allure. For example, they possess a kind of interiority to which the physical sciences remain oblivious. This interiority always eludes us in its entirety – exceeds what we can directly comprehend – and can be experienced only through metaphors that facilitate a transformative event of aesthetic attunement. He holds that in these moments of attunement, we can find ourselves addressed by a strong immediate sense of how we should respond. Referencing the *prima facie* rule-intuitionism of W. D. Ross, Magrini argues that in such moments we apprehend the nascent presence of a moral duty. For example, witnessing the results of the Exxon Valdez oil spill can awaken a visceral sense of a "grave and monumental injustice done to nature" and a duty to make amends as best we can (through restoration and changing future behaviour). This perhaps can be paralleled in the sphere of human ethics when Simone Weil (2005, p. 234) observed that "To know that this man who is hungry and thirsty really exists as much as I do – that is enough, the rest follows of itself". Proper perceptual attunement brings awareness of an ethical dimension.

Central to the argument here is that this sense of ethical challenge is invoked by the imperative force of the situation itself, not its rational or principled articulation that arises through a discursive assessment. Drawing on Alphonso Lingis's notion of "sonority" (as distinct from aurality), Magrini characterizes this ethical communication as an ever-renewed silent dialogue of address and rejoinder that is quickened by an underlying sense of the mortality and vulnerability to harm of all things, including the Earth as a whole. Amongst other things, one key element that participation in this sonority conveys is a sense of what belongs and what is alien to particular natural places. Following David Wood (2003), Magrini speaks of natural zones constituted by boundaries and thresholds, and that are constantly organically evolving through the rhythms and pulses of lived time. Experience of these zones establishes an existential sense of what rightly resides within them. A powerful – although negative – illustration of this is provided by the example of "seismic blasting" of the ocean floor in the search for oil deposits. Referencing Beth Wellington, Magrini describes how the process involves giant seismic airguns firing continuous blasts, sometimes for months on end with noise levels reaching 260 decibels that are greatly amplified underwater and can travel up to 2,500 miles. This has been found to cause harm not only to whales and dolphins that communicate by means of soundwaves in order to find food, avoid danger, and generally navigate, but also endangers sea turtles, fish, and zooplankton – the latter being a critical foundational plank of the ocean web (Wellington, 2011). And, of course, all of this is merely a prelude to the dirty and environmentally dangerous activity of offshore drilling. Here we are presented with a gross, but certainly not unique, example of a humanly instigated phenomenon that is both out of place – alien to a natural habitat – and that literally drowns out many imperatives emanating from nature. But not all, because experiencing the sound as alien and damaging itself constitutes an imperative.

Other illustrations abound. For example, there is the introduction of alien-invasive species (or "inns": invasive non-natural species as they have now been designated in the UK and elsewhere) into a habitat. These can seriously – sometimes completely – disrupt its original integrity so that communities of organisms that have evolved together and have achieved a delicate balance over long time periods of time are decimated. Often, full awareness of this is a product of extended and intimate acquaintanceship with the normal life and rhythms of this habitat (both causal and ontological) that imbues a sense of what is right, what is "healthy" or "wholesome", in this place and therefore what is being compromised. In this way, we might become aware of what is "out of character" in both the behaviour and the occurring of members of this biological community and the interplays of environing phenomena with which it interacts, and in which it subsists. Here, clearly, a strong sense of what is "natural" and its inherent worth are in play.

It might be thought that there are strong resonances between the view that is emerging here and that of Aldo Leopald's classical statement of environmental ethics: "A thing is right when it tends to preserve the integrity, stability and beauty of the biotic community. It is wrong when it tends otherwise" (Leopald, 1993, p. 382). By "biotic community", Leopald refers not only to animals and plants, but also soils and waters. In effect, all interrelating elements that constitute a living natural environment are accorded moral standing. While I have discussed some problems with this view elsewhere (Bonnett, 2004, pp. 109–110), taken in general terms it does articulate a position that places a proper emphasis on the integrity that arises from the holism of a natural place – the myriad, complex, and often subtle and delicate interrelationships that constitute it. These would have evolved (and continue to evolve) "naturally" within the biophysical milieu they create, and the logical space is provided for the idea of what belongs in this place and what would be alien. At base, there seems to be a sense of the sanctity of the natural forces and processes that are at work, and of a beauty that is the expression of scale and balance that is natural to that particular place in its own autonomy (yet, of course, in turn part of a still larger whole). Key to this view is the way that it blends morality, beauty, and natural causality into one phenomenon – or, more accurately, sees these arising as mutually sustaining facets of one underlying phenomenon: the integrity of a particular place. There is a sense, then, in which the view that is being developed in this book can be understood as a phenomenological reconstitution of Leopald's Land Ethic: a view that through its focus on how nature can be experienced itself – that is *for* itself – discloses the mode(s) of communication with nature through which this occurs.

Magrini quite rightly points out that given the significance and reticent nature of the dialogue in which these imperatives find a voice, it is important that we become alert to the enemies of this communication "in all the pernicious and insidious forms they assume". In addition to examples of the aforementioned kind, he refers us to the ways in which we attempt to make ourselves superior to the other (including nature), or subjugate it to our will, or shield our discourse from criticism – for example, by disingenuously filling time and space with the noise of irrelevant conflicting messages, "alternative narratives", or outright lies. In addition to these deliberate strategies, there is the numbing distraction of the sheer physical noise that the machinery and processes of modern technology can produce. These are all powerful obstacles to engaging in the silent dialogue with nature that is required. And perhaps, today, there is another source of distraction from, and distortion of, the full reality of nature that requires our attention – that of the burgeoning of digitalized experience. I will turn to this shortly. First, it is necessary to relate Magrini's view to the position that has emerged through previous chapters.

It seems to me that Magrini's account both resonates deeply with the account of nature's normativity developed in this book and helps to expand and refine it in some key respects. In particular, the *reticence* of nature's imperatives and the element of *ethical challenge* that they issue is underlined. They are a key constituent in the character of nature's voice. When we encounter an area of coastline decimated by an oil spill, it is silence that greets us. When we encounter what was once an area of forest now decimated by carpet logging, again it is silence that confronts us. In these examples, receptivity to nature's normative voice essentially is not a matter of perceiving physical sounds, but rather of intuiting the silent call of wrongness and a need (duty) to redress this. There is a perception of the destruction and its extent interfused with a sense of the moral harm that is being wrought. This latter, in turn, arises from a sense of the integrity of some natural locale or being that, to use the language of Chapter 2, arises through intuitively entering into the mutual sustaining interplay of the native occurring of things. Such participation is both sensuous and ethical, the one conditioning the other such that for the most part, when we are fully attuned and responding fittingly to nature's normativity – that is dwelling in nature by expressing its internal norms (as elucidated in Chapter 2) – the ethical remains entirely tacit. We are simply held in the sway of what nature invites.

Now, there have been those who, in the context of relating to nature, have placed the emphasis on knowing nature through sensitivity to, and learning to "read", the signs – sometimes subtle, and sometimes not so subtle – that arise in natural situations and surroundings. For example, this was one aspect of a rafting expedition on Tasmania's Franklin River, in which the participants sought to extend their lived experience of "all-inclusive nature, the more than human world", as agential and possessing "the potential to be considered as guide and co-teacher" (Ford and Blenkinsop, 2018). During their trip, the participants experienced a variety of elemental phenomena ranging from the roar and thunder of the rapids to subtle changes in temperature. An important part of reading the river consisted in understanding what these phenomena portended. To give one example:

> The first warming comes as a gentle whisper – a quiet puff rising up river and caressing exposed skin. . . . [F]or those who speak Franklin it is a moment of being put on notice. For a solitary breath of warmth might be nothing, but when the river starts to push sentences and paragraphs of the same language into your face that means rain is coming. And in this narrow trench of a canyon rain means rising water levels. And that means increasing danger and flooding.

They comment that one can give a meteorological explanation of this in terms of warm air from the deserts of Australia's interior meeting the moisture-laden cold air of the southern Indian Ocean, but suggest that "that all sounds like a post-rationalization of hard earned knowing" acquired by those who have spent a long time paddling the river.

To be sure, becoming intimately acquainted with the river in this way has a value. Direct experience of the elemental helps to reconnect – ground – us within nature in a way that modern society largely eschews, yet that is necessary for understanding the forces within which our existence is ultimately embedded. No doubt, it is also an enlivening experience for those involved. But although necessary, experiences had in this way are not sufficient for a full understanding of the reality of nature. In the light of the account that is being developed, these experiences, as described, seem still to focus on the surface of things – the physical phenomena and their causal connections – and as such they are not of themselves instantiations of nature's normative address. Certainly, they might speak of nature's integrity, which it has been argued is a fundamental condition of intuiting its normativity, but they stop short of that further degree of sensitivity that is required for the reality of nature to be more fully disclosed.

However, Ford and Blenkinsop give another example of their experience that moves closer to the idea of nature's normative address. While paddling along the river, suddenly they notice an animal sitting hunched at the water's edge its head turned towards them:

> In that moment its dark eyes meet mine.
>
> It does not move. I can feel it sensing us. "Wallaby" our guide says . . . do you know how unusual that is . . . to see a Wallaby here?"
>
> I feel petrified in the creature's gaze. The sensation of joy, the sensation of the traveller in a foreign land seeing something new gives way sharply to a deep sadness, almost a sense of shame. I "hear" the Franklin River's vulnerability, despite currently being protected, politically, from human interests.

Subsequently, the author reflects upon the quality and deeper resonances of his experience:

> For me, the wallaby had intimated something unexpected, unanticipated and had done so in a way that cannot be reduced to a simple refrain. When the wallaby spoke to me it was all at once. The communication was instant and seemingly contained within it a profound sense of melancholy, a warning of the future and maybe a moment of deep kinship. With its look the wallaby impressed itself on me, directing my mind.

He finds himself challenged to consider if his recounting of the experience is simply a personal projection, a kind of anthropomorphism:

> As the month pass, I begin to doubt my own experience, yet no matter how many times I replay the encounter it remains as vividly *participatory* as in the first instance.

I think that a number of important considerations arise from this account, some of which hark back to previous discussions about scientism and the disenchantment of nature. Why should we be tempted to dismiss the personal elements that reverberate throughout this account and the affective language through which it is expressed? What are the grounds for rejecting it as a portrayal of the reality of nature? Does not this account of the experience of the wallaby chime closely with what, in a human context, Levinas refers to as the "face" of a person, and what I have described as the living presence of a thing in its native occurring? If so, the objection to a scientisitic account here (again) is that it cannot do justice to the phenomena it purports to describe and that therefore its objectivity is spurious.

In a previous chapter, I discussed what, from the perspective of the full reality of nature, are the arbitrary prohibitions placed upon the language used to describe it by scientism. It seems to me that this discussion presents a powerful instantiation of this. If – as is fundamental to the intentionality thesis of consciousness previously set out – we participate in the reality of nature through an attentive consciousness to things themselves, and what *they* evoke, then experiences of the kind described in the preceding example become a legitimate, and indeed, central part of this reality. They need to be protected against the onslaught of the metaphysics of mastery and celebrated as a valid participation in, and insight into, the primordial enchantment of nature. They instantiate what a genuine encounter with nature in its wholeness involves, one that powerfully reconnects us to it. Indeed, on reflection, it is astonishing that we should ever have thought otherwise.

Digitalized experience

Today, it can appear superfluous to remark on the pervasiveness of digitalized Information and Communications Technology (ICT) in modern life. It has become integral to contemporary society – its institutions and processes – in so many ways. Along with water, gas, and electricity, broadband has become the "fourth utility" and "smart" devices are everywhere. The extent to which this digital diaspora is a good thing is a topic that far outruns the scope of this book, but there is an aspect of this phenomenon that increasingly is becoming central to its concerns: its effects on our relationship with nature and our ability to perceive it in its full reality – including, of course, our ability to receive and respond appropriately to its voice. Again, this, itself, is a very broad and complex topic, and the best that can be done here is to focus on some particular aspects that both are important in themselves and that can be taken to illustrate in concrete form something of the character of the broader issues that are at stake. To this end, I begin with an analysis of some features of experience that is mediated through personal computers. In keeping with other investigations in this book, the general approach will be phenomenological.

Typically, with personal computers, information – the world that it makes available – is organized so as to be conveniently available at the touch of a pad, few keys, or icons, and according to some algebraic logic that is energized not by the intrinsic significances of the contents that it locates – it is perfectly indifferent to these – but by consistency with the rules of its system. Clearly, at this level of interface, the logic of interaction and the character of our immersion in our environment is very different from what it is from when we are in direct contact with nature. In the former case, we are relatively disembodied, physically engaging by means of very limited finger movements structured by qwerty, touchpad, or screen and a very limited visual field. In the latter case, we are fully embodied: all the senses coming into play and stimulated in varying and unpredictable directions and ranges, and bodily movement is responsive, accommodating and potentially negotiating a far greater range of conditions. In these respects, as electronic information hunter-gatherers, we are positioned at a considerable distance from the textures of direct human experience and the kind of intimate reciprocity that I have argued to be inherent in experience of the natural world is fractured and occluded. In this sense, in computerized experience, we are radically reconstituted in our relationship with the world in a way that is far removed from – and is in danger of replacing – those habits of perception necessary to a full involvement in, and exploration of, the natural world. Indeed, it seems plausible to argue that in a world mediated by computers, the modality of our own being in the world is radically altered. We are no longer "there" in the same way.

For example, under the rubric of "technics of experience", Philip Payne (2006) has set out a number of cautions regarding the effects of ICT use on our ability to perceive nature, including the ways in which extensive computer use shapes, corrects, and reconstitutes both the body and our inner and social natures and the various natures with which it interacts. He asks:

> [I]n regularly using a computer how is the body physically positioned and socially arranged; what subsequent perceptions, interactions and relations with the wide range of "others" are structured; and how might such a computer "form of experience" shape constructions of nature?

He goes on to argue that

> [t]he "ontological" form of computer experience is primarily mentalistic, individualist and discursive/textual; sensorily static, repetitious and monotonous; socially constrained and environmentally constricted; geographically limited but spatially fluid; temporally concentrated but flexible.

While it is perilous to generalize extensively, certainly these are pretty straightforward features of a good deal of computer experience and clearly they have a powerful potential for alienating us from the kind of attentiveness that direct engagement with nature requires. Furthermore, the (often implicit) rhetoric that promotes the use of ICT and the ongoing everyday pressures to use it, in both in educational institutions and life outside, encourages participation in the assumption that what is presented in this medium is what is most significant – even primal – and occludes, if only by default, what is not readily represented within it: currently, everything that occurs through the senses other than sight and sound. And in computer-mediated experience, typically even sight and sound are of a peculiarly concentrated and limited kind, receptive only to what emanates from the device that lies immediately before it, the rest of the environment falling into near oblivion.

Clearly, in the light of such considerations, and with regard to how we are positioned in relation to the natural world, there is a need to examine how digital technology constructs and codifies both experience and the subjectivity of those who engage with it. And, while it is true that some of these considerations apply to technologies (such as print) that pre-date the ascendance of computers, the highly (yet sensuously narrowly focused) interactive and engaging qualities of the latter, their calculative, organizational, and presentational power, and their protean character in terms of software and plug-ins that allow them to thrive in almost any educational habitat (Scrimshaw, 1989) serve to intensify them. For example, there is an important sense in which in much computer use – particularly in the case of virtual realities and games – our sensibility is heavily conditioned in a further way compared to the way that it is in reading books. The printed word depicts a reality by engaging the imagination that is free to draw upon a wealth of experience deriving from prior direct sensory encounters in our lived experience, including that of the natural world. In comparison, the "realities" of computerized experience within which the participant engages are heavily pre-selective in this regard (and pre-structured according to someone else's taste), and the free play of imagination is therefore significantly curtailed. This represents a further degree of attenuation of, and alienation from, the reality of the native occurring of things in nature. And whatever the details, in general terms the mutually supporting phenomena of the growth of scientism and the growth of ICT that are such powerful elements in late-modern society, can be seen as examples of the playing out of a metaphysics of mastery. Both are complicit in an ordering of everything so as to make it available to, and responsive to, a sensually restricted approach to reality that is convenient to, and embodies, the purposes of manipulation and mastery. Presently, I will argue that in their insulation from the full sensory reality of nature, both facilitate a frame of mind in which unsustainability can become normalized because the direct experience of what is afoot in nature and that can evoke and sustain a grounded environmental concern has become downgraded and attenuated. In Chapter 3, I referred to some ways in which the derogation of nature has been reflected and reinforced by powerful strands of thinking in academia. It is now time to consider how a reaffirmation of the kind of truth that direct experience of nature in its occurring supplies could affect our understanding of education and contribute to its ecologization.

Ecological truth

If, in contra-distinction to Rortyian pragmatism, we allow truth to be more than merely some empty compliment that we pay to propositions to enable us to get on, we can see that it is both central to human sensibility and that, because of its self-arising quality, the truths of nature elicit kinds of knowing that are dialogical and intuitive. As previously argued, they involve a sensing of things in their particularity that embody and express agency and value, motion and mystery. Hence, they require modes of attentiveness that are open, multi-sensory, bodied and affective as well as cerebral.

Given this account, one of the considerations that the issue of knowing nature raises is the character of truth and the conceptions of it that inform/should inform education. There are many facets to this, but one key point is an invitation to consider the possibility that the focus on receptivity to the manifoldness of things themselves can foreground an *ontological* notion of truth: the significance of the idea of "being in truth". This is to say that it can draw attention to the quality of our living relationship with the thing known, the way that we *are* towards it.

There are some clear resonances here with Heidegger's (1962, Sect 44, 1978) distinction between truth as correctness and truth as disclosure, in that these involve fundamentally differing stances towards the thing known, and in the latter case, involves a participatory "being-with". In the environmental context, this focus on the quality of the relationship rather than on the thing as an object of knowledge could be seen to resonate with David Abram's claim:

> Ecologically considered, it is not primarily our verbal statements that are "true" or "false", but rather the kind of relations that we sustain with the rest of nature. A human community that lives in a mutually beneficial relation with the surrounding earth is a community, we might say, that lives in truth. The ways of speaking common to that community – the claims and beliefs that enable such reciprocity to perpetuate itself – are, in this important sense, *true*. . . . A civilization that relentlessly destroys the living land it inhabits is not well acquainted with *truth*, regardless of how many supposed facts it has amassed regarding the calculable properties of its world.
>
> *(Abram, 1997, p. 264)*

Of course, Abram's extending of the idea of truth in this way that connects it to what is "mutually beneficial" clearly presupposes notions of human well-being and the Earth's well-being. Both notions are not without their problems, but, as I have tried to demonstrate in earlier chapters, through a sensitivity to intrinsic value and normativity in nature and the way in which sustainability is integral to human consciousness, there are defensible ways of understanding these terms and of bringing them into relationship.

In addition, from a Western perspective, Abram's view opens the possibility of truth as correctness of verbal statements not simply being distinguishable from "living in truth", but coming into conflict with it. For example, the animism that characterizes the views of many indigenous peoples who have lived in harmony with their environment over long periods (and therefore "lived in truth") is likely to be seen as clearly incorrect as a statement about the real (scientific) character of physical reality. Surely, an indigenous culture is simply wrong if it believes that stones are sensate and possess memories in a manner analogous to the way that humans do.

Circumventing the issue (after postmodernism) of whether this assertion is an example of Western scientific hegemony – that is, *scientism* – there is another question that can be posed: even with regard to phenomena that, from a Western perspective, it might be felt properly fall within the domain of science, is it possible that something that is scientifically false can nonetheless express a greater truth? Namely, to reveal a quality of relationship with the natural world that (a) in its own way and through its own metaphors expresses something of worth, such as an intimate simpatico with particular natural phenomena that allows them to presence more fully; (b) reveals starkly the partiality of scientific truth and its limitations in providing a portrayal of reality that should infuse our life-world.

Scientism has extensive implications for how we understand and live truth because of the kind of reality it generalizes beyond the domain of science and into the life-world as a whole. In modern times, increasingly we have become accustomed to thinking of empirical truths as autonomous in the sense that statements about the world can be correct irrespective of, say, their moral or aesthetic implications or the consequences of acting in accord with them (however, that is to be determined) for the world's well-being. This is reflected in the idea of the traditional fact-value distinction that claims that one cannot derive an "ought" from an "is" because when empirical facts are stated, it is always possible to ask whether they have the moral implications attributed to them, or whether the situation they describe is what it ought to be. This is taken to demonstrate that moral or other value is not *internal* to an empirical fact about the world. For example, from the fact that, say, a person has a particular talent – from that fact alone – we are not warranted to infer that they should exercise it. From the fact alone, that a forest is ancient or a river has flowed freely from time immemorial, it does not follow that the forest should not be logged or the river dammed. In each case, a further value claim is required and this can only be derived from a world of non-empirical propositions whose claims, because in principle they cannot be supported through purely empirical observation, are often regarded as subjective and therefore problematic.

But this only holds true if we first posit or presuppose as primal a neutral world such as the mechanical world of the physical sciences. Then, even if it were possible at specific points and with regard to particular aspects (e.g., in specific moments of participation with nature in place-making) to discern intimations of well-being for nature or humanity, these would have to be understood as (mere) human projections upon the world. However, if, as has been argued in previous chapters, the world is understood as inherently normative, this changes. Truth and well-being become connected; the latter becomes integral to truth. Propositions and actions that contradict such normativity are no longer true to the world. They go against the flow of the world and therefore do not express it and cannot represent it: they are false to it – reflecting, perhaps, only the machinations of an insulated human will. It is in this context, when a wholehearted participation in the flow of things is broken, that purely human-authored, detached purposes and principles (rational or otherwise) arise and assume the countenance of sources of morality. In this moment, the knowledge of abstract generalization severed from the knowledge of intimate engagement holds forth and the products of alienation come to gain authority over, and to extinguish, experiences of acquaintanceship.

In this moment, too, the possibility arises of such unrooted knowledge eddying off from full reality and becoming available for service to the metaphysics of mastery, and of the natural environment being encountered as pure resource. Yet it is receptivity to the flow of value, affection, that binds us to the world – indeed, that, as Max Scheler (1980, p. 109) expressed in his notion of "value-ception", first places us in it so that we can properly perceive it. This is because (a) values are an integral part of the lived world and stream off to us in our immediate experience of it; (b) in being receptive to intrinsic value in the other and hence receptive to being that lies beyond our self-given instrumental purposes, we are open to the fullness of things themselves in the inherent mystery of their sheer "suchness" as they proceed on their own ultimately ineffable trajectories.

Consistent with previous discussion, the argument here is that it is the self-assertive frame of mind that remains oblivious to such intimate felt connection that lies at the kernel of our current environmental crisis – is its hidden centre – and therefore must be exposed and confronted in all its myriad expressions if substantive progress is to be made in addressing our environmental crisis. Perhaps its key form of expression is the pre-specification that pervades modern life from modular window frames to modular curricula. To take the latter: the kind of knowledge and learning that can be pre-specified and "delivered" is about as far away as it is possible to get from the kind of knowledge and learning that intimate acquaintanceship with nature evokes. The living presence of natural things previously described simply cannot be captured in pre-specified items of knowledge, and indeed, because of its intense particularity is not fully articulable (Bonnett, 2004, p. 93). Acknowledgement of this transforms the curriculum from one where pre-specified material is transmitted and its reception is checked by standardized tests to one in which what is to be learnt arises through the fluid and emplaced interplay of pupil and teacher engagement with what presents itself (in its mystery) in the course of a genuinely receptive-responsive enquiry. It is, therefore, a curriculum (and culture) of emergent engagements, whose unity is not the result of pre-formed disciplinary, interdisciplinary, or utilitarian connections determined by those distant from particular sites of learning, but the result of an evolving interplay of felt demands arising in the course of receptive participation in the draw of the known and the as yet unknown: the never to be fully known.

On this account, to impose a detailed pre-specified structure on learning is to cabin the human spirit and to efface the many-sidedness of people and things; it is to condemn the growth of organic learning to be generalized out of existence. In this way, many of the negatives at large in a scientistic culture come to reverberate in a particularly concentrated form in the educational space in which children are invited to participate, and are compelled to inhabit – exhibiting themselves as powerfully in the culture of the school as in the overt curriculum (Bonnett, 2010).

The "greater whole" in environmental discourse

Within the sphere of environmental debate, there are a number of views that, in different ways, place heavy emphasis on conceiving humankind primarily as a part of some greater whole or system. For example, this is a feature of "deep green" and associated eco-centric philosophies such as that developed in Freya Matthews' *The Ecological Self* (1994). In this influential book, she argues that as each individual is nested in a vast sustaining system whose flourishing is a *sine qua non* of their own flourishing, we should see the universe (which she refers to as the "ecocosm") as our "extended self" with which we therefore identify and love as ourselves, and of which our individual self is tributary – a localization in which the ecocosm achieves self-awareness. And although rather different in other respects, this emphasis on the primacy of the whole is a feature shared with Paul Taylor's *Respect for Nature* (1986) that emphasizes our oneness with "the great community of life" in which there is a fundamental interdependence and therefore equality between all its members. It is also emphasized by those who adopt a radical systems thinking approach to environmental issues and education.

More generally, there is of course a long tradition of emphasizing the importance of acknowledging our existence within some greater whole. For example, A. N. Whitehead wrote:

> That which endures is limited, obstructive, intolerant, infecting its environment with its own aspects. But it is not self-sufficient. The aspects of all things enter into its very nature. It is only itself as drawing together into its own limitation the larger whole in which it finds itself. Conversely it is only itself by lending its aspects to this same environment in which it finds itself.
>
> *(Whitehead, 1925, p. 94)*

In the context of evolutionary theory, W. H. Murdy made the point:

> In order to influence evolution in wise and responsible ways, we must strive for an ever fuller understanding of our relationship to greater wholes – society, nature, and ultimately to the primary source of order and value in the world. Personal identification with greater wholes is essential to the discovery of our own wholeness.
>
> *(Murdy, 1975, pp. 1168–1172)*

Clearly, invitations to reject or bridge the divide between humanity and nature that characterizes traditional occidental thinking offer an important perspective on addressing our current environmental predicament. The history of exploiting nature in ways that have deleterious consequences is often attributed in part at least to ignoring the reciprocity that exists between human well-being and certain states of nature. For some, the realization of human-induced climate change and its anticipated consequences has been a rude awakening in this regard. Hence, undoubtedly, it is important to raise the level of our appreciation of the relatedness and interdependence that exists between ourselves and the environment. We need thoroughly to understand the character of this interrelatedness and to shape our actions in ways that truly reflect this understanding. Amongst other things – and if foregoing argument in this book is correct – it is important to bear in mind that this relationship is not now simply bio-physical, but also *meta*physical.

In developing this theme (and as previously has been noted), a key consideration is that of what we count as knowledge, and the relative status we accord to different kinds of knowledge. Of particular relevance here will be an examination of the motives that energize different kinds of knowledge, but also our knowledge attributions more generally – for example, the tendency to elevate to the status of knowledge primarily those descriptions and interpretations that give power. This latter is illustrated by a strong tendency in education to value – as the terminology itself makes clear – so-called "higher order" thinking skills and knowledge. This refers to the kind of thinking and knowledge that analyses and synthesizes "lower order" particulars into categories, hypotheses, and theories that enable greater explanatory power and the more extensive utilization of what is known. Here the being of each particularity is made subservient to achieving some greater purpose. Hence, knowledge of key concepts such as "energy" and "mass" or "evolution" and "natural selection" that can be seen to organize thinking within the various research disciplines are considered to be superior to what some indigenous culture might know first-hand about its locality, the detailed habit of a familiar animal, or some particular recent occurrence. Yet this more concrete, particular, and emplaced knowledge is precisely the kind that draws us into a closer personal relationship with nature.

In the light of considerations of this kind – and in tandem with the discussion of an ontological notion of truth – it will be argued that there is a need significantly to transform the ideas of knowledge that inform conventional subject-based school curricula. In addition, and playing into this area of debate, it is necessary to express two caveats concerning the focus on the "greater whole" in understanding ourselves and our situation. First, there is a concern about the way that this greater whole is characterized – for example, the impact of some of the metaphors used upon our conception of ourselves and the character of our responsibility towards nature and the environment. Second, there are issues concerning the degree of inclusiveness that this point of reference is assumed to possess as an explanatory notion. As they provide a useful preface to exploring issues of knowledge transformation that an ecologized education requires, I will take these reservations first. I will illustrate them by reference to some lines of thinking developed by the anthropologist Gregory Bateson, whose prototypal ideas first published some 40 years ago continue to express much that is characteristic of systems thinking approaches that have become influential in much current discussion (see, for example, Bowers, 2011; Sterling, 2017). In several respects, what follows can be seen as an amplification of the preliminary discussion of systems thinking broached in Chapter 2.

In his influential *Steps to an Ecology of Mind* (2000 edition), Bateson makes a range of points concerning our relationship to a greater whole that, taken in themselves, possess a high degree of plausibility – and importance. For example, he makes the point that as parts of the whole, we can never be in a position to see the whole. Consciousness is always selective, working by a systematic sampling of the events and processes of the body and of what goes on outside according to its purposes. This means that it lacks *systemic* wisdom. To those who might reply that this has worked well enough so far, Bateson draws attention to the addition of modern technology to the old system. He holds that consciousness is now empowered to upset the balances of the body, society, and the biological world – to "wreck" the environment (p. 452). Hence, this lack of systemic wisdom is now pathological. In an overarching system portrayed as one whose elements are set to grow exponentially and are only held in check by resources and the curtailments that one growth career exerts upon another through various feedback mechanisms, lack of systemic wisdom is always punished. So situated, our actions can have extensive consequences for the environment that we cannot predict, and over which, once set in train, we might have little control. There is a good chance that climate change will turn out to be an example of this, but such is the complexity and temporal and spatial extent of natural systems that it is likely that many other, as yet unrecognized, collateral effects of human actions are working their way out, and of whose "tipping points" we are entirely ignorant. Hence, it is essential that our narrow purposive view somehow be corrected. Bateson goes on to point out that this lack of systemic wisdom is particularly problematic in the face of two further considerations:

1 Man's (*sic*) habit in the face of a changing variable is to change his environment rather than himself;
2 The way that today the social scene is characterized by a large number of self-maximizing entities, such as trusts, companies, political parties, unions, commercial and financial agencies, nations, etc., that have something like the status of (and therefore legal rights of) "persons", but that are precisely *not* persons and are not even aggregates of whole persons. They are aggregates of *parts* of persons – in particular precisely the highly conscious narrowly purposive parts. Thus, these social entities contribute to isolating conscious purpose from many corrective processes that might come out of less conscious parts of the mind (*ibid.* p. 452).

Certainly, these are pertinent points – and not least in their implicit acknowledgement of aspects of persons that apprehend reality in less purpose-driven ways, but that have become peripheralized by an overweening instrumentality. This is entirely consistent with the idea of a prevailing metaphysics of mastery that has been a central theme of the current work. But it seems to me that a serious problem arises with the underlying way in which Bateson goes on to construe the greater whole and the character of wisdom. As his thinking on this matter has produced a template that has influenced many other thinkers in the area, I will pursue it a little further here. Basically, on his account, the picture is one of the greater whole conceived as a vast cybernetic system – a "self-corrective information feedback system". And the individual mind is portrayed as a variable localized part of this system. Here we have the beginnings of a radical dissolution of the traditional idea of individual centres of consciousness – distinguishable selves that possess their own interior integrity and that persist through time. As Bateson puts it:

> "My" mind – delimitation of an individual mind must always depend upon what phenomena we wish to explain: Obviously there are lots of message pathways outside the skin, and these and the messages which they carry must be included as part of the mental system whenever they are relevant.
>
> (ibid. *p. 464)*

This notion of a "message pathway" is employed to override the traditional distinction between the mental and the physical; all is pervaded by "Mind". As for the notion of an individual consciousness:

> But what of "me"? Suppose I am a blind man, and I use a stick. I go tap, tap, tap. Where do *I* start? Is my mental system bounded at the handle of the stick? Is it bounded by my skin? Does it start halfway up the stick? Does it start at the tip of the stick? But these are nonsense questions. The stick is a pathway along which transforms of difference are being transmitted. The way to delineate the system is to draw the limiting line in such a way that you do not cut any of these pathways in ways that leave things inexplicable. If what you are trying to explain is a given piece of behaviour, such as the locomotion of the blind man, then, for this purpose, you will need, the street, the stick, the man; the street, the stick, and so on, round and round.
>
> (ibid. *p. 465)*

Now a need to acknowledge more fully that we exist in relationship to an environment has been a central theme of this book, but here we are invited to subscribe to a picture of mind as synonymous with a cybernetic system – as Bateson puts it: the relevant "total information-processing, trial-and-error completing unit". Furthermore, on Bateson's account, we are required to recognize that within Mind in the widest sense, there will be a hierarchy of sub-systems, any one of which we can call an individual mind (p. 466). For what Bateson dubs a "cybernetic epistemology", the individual mind is immanent not only in the body, but also in pathways and messages outside the body. In addition, he speaks of a larger Mind of which the individual mind is only a sub-system and whose identity varies with the phenomena to be explained. Bateson suggests that this larger Mind is comparable to God (he supposes that for some maybe it *is* God), but is now strictly immanent in rather than transcendent to "the total interconnected social system and planetary ecology" (p. 467). He suggests that just as Freud expanded the concept of mind inwards to include an unconscious, this view expands it outwards. Both reduce the scope of the conscious self. For Bateson, such a perspective involves a humility tempered by the dignity of being part of something much bigger: a part, as it were, of "God". And it requires a new way of thinking that dissolves the traditional – "pre-cybernetic" – notion of the self where the "Myself" is an excessively concrete object different from the rest of "mind" (p. 468).

At first blush, it might appear that this portrayal of mind/consciousness could be interpreted as a full-blown version of the environmental consciousness that was elaborated in Chapter 4. There, too, consciousness was considered to be not self-contained, but to be always reaching out beyond itself. Indeed, it was conceived to be part constituted by transcendent objects that exist in the world – a public shared world. But there are a number of crucial differences between this account and Bateson's cybernetic model. First, what Bateson terms the "Myself" is not dissipated through myriad "message pathways", but survives as a distinguishable being that has an interiority and evolving identity. Its experiences have the quality of "mineness" in that there is an important sense in which they are felt to belong to it and, critically, it recognizes a responsibility for its actions. Both of these features speak of an individual mind that possesses a degree of persistence and stability, and whose identity is not to be regarded as radically reconstituted in response to the arising of different phenomena to be explained. Second, environmental consciousness is not conceived primarily in terms of information flow, but rather in terms of receptivity to the occurring of things in which personal mood and emotion are evoked and fitting response is elicited or contemplated. To be sure, this is not to deny that there is a flow of what one might term "information" in the sense that significances are communicated and created in the flow of intentionality, but for reasons that I will set out presently I think that the way in which Bateson uses the term "information" has connotations that are antipathetic to the idea of being as occurring and that convey a highly reductionist view of the natural world – a view, indeed, that contributes to an unwarranted disenchantment.

To return to Bateson's account, it seems to me that a range of general problems attach to his cybernetic model of world and self, and the epistemology that flows from it. Here I will focus on some that have a direct bearing on the concerns of this book: our understanding of ourselves in relation to nature. I will begin by raising, somewhat briefly, two closely interrelated reservations before considering a further one in more depth.

First, there is the seemingly unavoidable characterization of mind as a computer, simply a "sub-system", a "total information-processing, trial-and-error completing unit". Bateson constantly speaks in these terms, and with no sense of any loss involved to our understanding of either self or world. Yet in the context of our relationship to nature and the environment, the determinism that this metaphor connotes is of particular concern because it ultimately undermines ideas of personal agency and conscience. If, philosophically, these are "old chestnuts", in the current context they remain "hot" ones, for it is hard to reconcile such determinism with calls for responsible behaviour towards the environment – or indeed with any *personal* understanding of the environment at all, if (as I think it must be) such understanding is taken to occur in the context of practices undertaken by responsible agents capable of mortality and for whom, therefore, to follow Heidegger, "their own being is an issue". In other words, the computer metaphor radically revises the whole landscape of human understanding, dismissing some of its most salient features with no recognition of the losses incurred. Some of these will be visited presently, but to reiterate here, the intentionality of consciousness as developed in this book is not simply a process of information flows, but an openness to what is radically other and that includes elements whose essential qualities are distorted or impoverished when construed in these terms.

The second, related, reservation concerns the characterization of the mind as a series of transient localizations of the greater system. In the resultant blurring/attenuation/dissolution of self–other (nature) boundaries, what happens to (a) our sense of an enduring personal identity and (b) our sense of the otherness of nature and hence the character of our respect for it? As I have indicated, it is not my intention to develop these reservations in any depth here. However, by way of setting a backcloth to issues that presently I will pursue more fully, it is worth pointing out that the analyses of nature that I have given hinge on its autonomy from human intention and its intrinsic mystery, and analyses of human consciousness (or its equivalent) that centre on its distinction as the place where things show up with a certain kind of significance and degree of stability, are likely to be seriously discomforted by Bateson's cybernetic characterizations. And while this might (or might not) be his intention, his is a view whose large consequences are hugely counter-intuitive and need to be carefully examined. Again, this is something that Bateson conspicuously fails to do and some of the salience of the issues thus left unvoiced will appear in the analyses that follow.

I turn now to a reservation that I wish to develop in more depth: the invitation to construe nature primarily as an information system or flow and that understanding it in these terms is a proper basis for the systemic wisdom required for environmental decision-making.

Nature as information system

The issue at stake here is not that there are not certain insights to be gained from construing nature as a cybernetic system, but that these should be recognized as limited and as serving strictly limited purposes, rather than being set up as the most fundamental way of understanding nature, providing therefore the most appropriate orientation towards it. For example – and to put it rather crudely in the first instance – why privilege seeing a pond or hedgerow as an information system? Clearly, on the view endorsed in this book, this would be to elevate what can equally well appear as a scientisitic demotion of life-world experiences and attachments that can be a seminal source of care for nature.

It seems to me that a central aspect of the problem here is revealed by an analysis of the idea of information (as distinct from knowledge more generally). "Information" as now frequently spoken of belongs to an instrumental frame of mind. It is knowledge that has been processed to meet particular requirements: perhaps to answer a question or solve a problem. It is sought or given where people wish to, or are considered to need to, know specific things for specific purposes. This is to say that it is a form of data organized for human purposes and therefore it occurs at a remove from the raw reality of the native occurring of things in experience. Indeed, Bateson's locating information in the context of information-processing feedback systems suggests superordinate elements of quantification and calculation that lend weight to this analysis. And in being tailored so as to be appropriate to a pre-specified purpose, such information is revealed as necessarily a highly selective abstraction that effaces many "irrelevant" facets of this occurring, closing it down and reifying it. In all, the term "information" is revealed as a semantic instrument of the metaphysics of mastery and as such, if somehow taken to encapsulate mind/consciousness and the nature of experience severely impoverishes our life-world. Knowledge, wisdom, and fulfilment in our relationship with nature grow precisely in this fluid life-world of spontaneous occurrences; they are the achievement of an intimacy that develops dialogically between experiences over time that are unfettered by external structures of pre-specification. They are redolent with feeling and emotion, subtle and inextricable associations – and as has been previously argued – mutual anticipations – that constitute a form of sensibility that far outruns and surpasses what ideas of information flow can legitimately convey.

In this context, we are returned to the important issue of how we conceive the constitution of things in nature and our relationship to them, and the radical implications that this has for what counts as authentic knowledge of them. Here the issue of the status accorded to different kinds of knowledge arises – particularly with regard to the twin hegemonies of intellectualism and abstraction that in various guises have received such a fillip from the overt successes of science. For example, in the context of requiring global solutions to environmental problems, it is tempting to seek, and to rely upon, what might be termed "universal" or "international" knowledge – knowledge generated and configured so as to be applicable to a wide range of situations by being existentially rooted in none of them. Indeed, it invariably shares many of the generic characteristics of information as already described, being organized into packages of theoretical models that can be distributed from some high status, essentially free-floating, point of origin and imposed with only peripheral modifications upon situations that in many highly significant ways are uniquely locally conditioned, but that are now perceived as systems of the same generic kind, and to which they become assimilated.

Maurice Merleau-Ponty is one of a number of thinkers who developed a perspective that is diametrically opposed to this kind of scientific systems approach and who illustrates the significance of reasserting the primacy of an individual's life-world over scientific abstraction. Returning to the question of things in nature, in *Phenomenology of Perception* (1962, Part 2, Ch. 3), Merleau-Ponty asks: what is it that constitutes the constancy of the thing – its reality? He criticizes the scientific approach that sees a thing as defined in terms of its location and dimensions within an *a priori* spatial-temporal framework and whose appearances are understood according to laws that relate them to how it would be perceived under a set of standard conditions – as, say when viewed from a certain angle, a diamond shape is interpreted as "really" a square, or a colour is seen as, say, "really" black when because of lighting conditions it appears as grey by being intellectually interpreted – calculated – by reference to a set of "pure" properties. In this view, "Reality is not a crucial appearance underlying the rest, it is the framework of relations with which all appearances tally" (p. 300).

For Merleau-Ponty, this is subject to two deficiencies: first, it is phenomenologically inaccurate; second, it begs the question as to how we come to have the idea of a determinate object in the first place. He observes that:

> Perceptual consciousness does not give us perception as a body of organised knowledge, or the size and shape of the object as laws; the numerical specifications of science retrace the outline of a constitution of the world which is already realised before [such] shape and size come into being.
>
> (ibid. *p. 301)*

Merleau-Ponty makes the point that science takes the results of this pre-scientific experience for granted. Far from its being the case that the thing is reducible to scientifically specifiable constant relationships, the latter are based upon the self-evidence of the thing as given to us in perceptions that we take to typify it. As he puts it:

> For science and objective thought, an apparently small object seen a hundred yards away is indistinguishable from the same objects seen ten yards away at a greater angle, and the object is nothing but the constant product of the distance multiplied by the apparent size. But for me the perceiver, the object a hundred yards away is not real and present in the sense in which it is at ten yards, and I identify the object in all its positions, at all its distances, in all appearances, in so far as all the perspectives converge towards the perception which I obtain at a certain distance and with a certain typical orientation.
>
> (ibid. *p. 302)*

It is this privileged perception that ensures the unity of the perceptual process by acting as the reference for all other appearances. Hence, he argues that:

> For each object, as for each picture in an art gallery, there is an optimum distance from which it requires to be seen, a direction viewed from which it vouchsafes most of itself. . . . [A] living body, seen at too close quarters, and divorced from any background from which it can stand out, is no longer a living body, but a mass of matter as outlandish as a lunar landscape, as can be appreciated by inspecting a segment of skin through a magnifying glass.

> Again, seen from too great a distance, the body loses its living value, and is seen simply as a puppet or automaton. The living body *itself* appears when its microstructure is neither excessively nor insufficiently visible, and this moment equally determines its real size and shape.
>
> (ibid. *p. 302)*

It follows that in order to perceive something, the body must be stationed before it and in seeking to see it better, it adopts a certain position. In this sense, appearances are always enveloped for me in a certain bodily attitude. This leads to a very important conclusion:

> The constancy of forms and sizes in perception is therefore not an intellectual function, but an existential one, which means that it has to be related to the pre-logical act by which the subject takes up his place in the world.
>
> (ibid. *p. 303 footnote)*

At this point, we arrive at the kernel of the matter in terms of what systems thinking occludes, and that has been frequently affirmed throughout the current work: the centrality of an emplaced subjective being in all perception. And thus, while any perception of a thing, a shape or a size etc., as real – that is, any perceptual constancy – refers back to the positing of a world and in some sense a system of experience, this latter is

> not arrayed before me as if I were God, it is lived by me from a certain point of view; I am not the spectator, I am involved, and it is my involvement in a point of view which makes possible both the finiteness of my perception and its opening out upon the complete world as a horizon of every perception.
>
> (ibid. *p. 304)*

In the light of this ongoing emphasis on the role of personal bodily situated participation in the primordial occurring of things and their constitution as individual entities in the world, and on the shortcomings of an understanding that seeks to "array all before us" in some purely objective way, the true character of systemic wisdom now can be identified. It can perhaps be summarized in the following way.

Systemic wisdom

Over the course of history, there have been many versions of "the greater whole". Amongst the more dominant modern conceptions, perhaps the most ubiquitous is that of a law-governed causal network employed by classical science and related to this, the network of bio-physical sustainability employed by scientific ecology. In addition, more recently, and as discussed previously, the systems thinking approach in which the environment as a whole is conceived as an information feedback system has gained some prominence. The key point about all of these conceptions is that they represent highly *discursive* forms of structure. They involve a systematization achieved through the imposition of a super-ordinate abstract theoretical framework that, in accordance with its own logic, delegates to each aspect of experience its character, place, and significance. The upshot of this is that beneath the appearance of a holism here lies a kind of atomism. This occurs in the sense that the fluid continuity of mutual sustaining and anticipation essential to the native occurring of things in nature – and the participation of an individual embodied subjectivity in this – has been disrupted. This living milieu is reductively broken up and ossified so that component elements and processes can be objectively specified. Accounts of this kind, through their inattentiveness to the primary character of the native occurring of things (and in particular, the "self-arising" quality of natural things), unavoidably do violence to the phenomena that they purport to describe.

So, what might an alternative account look like? Enlarging on the theme of the preceding comment, access to a non-discursive "whole" is facilitated by sensing the transcendent that is immanent in the particular and the infinite interplay of coming into presence and withdrawing in the occurring of things. It requires that we travel with the self-arising flow of the gathering and the dissolution of things that yet possesses a pressing reality. The self as part of a greater whole, in the sense of participating in its standing forth, engages most authentically with that greater whole not chiefly by abstract ratiocination (although, appropriately conditioned, there is a place for this) but through the ongoing simpatico of its immediate dealings with the individuals it encounters; its attunement to its own and their own participation in the infinite interplay of native occurring as it spreads out in all directions to the horizon and beyond.

Furthermore, it is clear that the preceding account suggests an understanding of the normative significance of the environment for education in a fuller sense than is often recognized. The account makes it clear that what is at issue is not simply or primarily the matter of addressing our environmental predicament as gauged against the kinds of degradation to which attention is drawn by ideas of climate change, pollution, and species extinction – important as undoubtedly they are. Rather, what becomes central is the quality of our individual being in a world that is constituted by an ever extending and subtly nuanced interplay of significances and mutual anticipations that have sensuous as well as cerebral, aesthetic and ethical as well as prudential, dimensions. At the heart of this mode of being is a caring that is open to non-anthropocentric impulses – a caring that lies in a recognition of the reality of the other and the responsibility for non-interference that such recognition incurs. It invites a celebration of alterity, of the sheer existential autonomy of non-human nature, and of the nobility of that which is truly itself. This necessarily checks the rush to action and the drive for mastery. A whole that is not regarded essentially as a resource and that therefore is not in need of definition, organization, and manipulation can be accepted in its fluidity and mystery. So apprehended, it is allowed to flourish in a space in which it can gift inspiration. Only by listening for the call of what is other can we begin to receive the norms that imbue the places (neighbourhoods) we inhabit and so receive intimations of how our existence is in interplay with all else. In this way, a sensing of a greater whole that is neither totalizing nor atomizing, but that embraces the ever-changing cadences of the places – the populated (but not simply peopled) neighbourhoods – that constitute it can occur. Herein lies the foundation of environmental wisdom.

7

ECOLOGIZING EDUCATION

Introduction

The arguments concerning the primordial phenomenological character of nature as a native occurring of things and the kind of consciousness to which it relates have led to the examination of a wide range of issues. As was noted when they were discussed, many of these issues have indicated the need to reappraise philosophically some key and long-standing conceptions that currently inform much thinking about the environment and environmental policy. Indeed, cumulatively, the import of foregoing chapters points to the need for deep cultural change. Undoubtedly, this is easy to say and difficult to achieve. There are many complexities, and at best any such change is likely to be a gradual process involving education in a broad sense. It is now time to turn explicitly to the effect of these revisions on how we should think about education and what might be involved in ecologizing it. Arguments that have identified a need to combat highly pervasive and deeply embedded cultural influences such as those concerning a prevailing metaphysics of mastery, or that call for modification of some influential ideas of consciousness and selfhood – ideas that stand at the heart of much conventional educational thinking – presumably will have extensive implications for ecologizing education. For example, they raise broad and highly interrelated questions of the following kind:

> How should we confront the hold of the metaphysics of mastery and scientism in education and what might this involve?
>
> What are the implications for education of seeking to develop emplaced selfhood and environmental consciousness?
>
> What are the implications for the moral culture of education and its underlying spiritual orientation of recognizing nature's voice and moral standing?

On the view defended in this work, addressing these questions would be foundational to addressing our current disastrous environmental situation and to facilitating a better and more authentic relationship with the natural world than that which currently prevails. On the basis of our responses to these questions, it should be possible to discern the kinds of content, practices, and experiences that an education devoted to this end would involve. Hence, this chapter will pursue the strategy of revisiting and amplifying key aspects of the philosophical underpinnings noted previously as a prelude to considering more detailed and specific ideas for educational practice that arise from environmental concern so conceived.

However, before taking this further, there is an important point that I wish to clarify concerning the relationship between the arguments that I have expounded and the place of science in education. This clarification is necessary because traditionally science (often along with geography) has been the curriculum home for much environmental education. In addition, it is largely on the basis of scientific evidence and the authority of the voice of scientists that environmental problems of pollution and climate change have come to be taken more seriously at a political level. First, it needs to be emphasized that throughout the arguments that I have mounted, it has been scientism and not science that has been the focus of criticism, and that the arguments have been offered in the spirit of seeking to identify and to redress an imbalance in perceptions about how we know nature and what approaches are most relevant to addressing environmental problems. In part they amount to a set of arguments against a response to our environmental predicament that is founded on an overweening anthropocentric stance and that consists in seeking and relying upon technological fixes. While it is not being argued that technological solutions have no place in our response – for certainly, they do – the position that I have developed makes it clear that these and the understandings of the environment that initiate them are neither the whole of the story, nor, ultimately, the most fundamental part of the story. This is partly because what lies at the heart of my account is that our current environmental predicament is a crisis not simply of a lack of empirical knowledge and expertise that would enhance our power to control events, but rather one of our underlying attitudes towards the world in which we are embedded. Therefore, it is a crisis not only of our physical survival, but of our spiritual being – that is, our felt intuitive understanding of the kind of beings that we are and how we should relate to everything around us.

This assertion raises issues that are as much matters of human emotion and morality as they are of the intellect and scientific fact. Indeed, it suggests that the crisis that now needs to be recognized is that of our whole mode of sensibility – the way that we are towards the world and are capable of receiving it. Furthermore, ultimately if this ontological/spiritual dimension is ignored, our physical survival is imperilled, for it is held that the metaphysics of mastery lies at the centre of the human behaviours that have brought us to our present pass and if allowed to continue unabated (notwithstanding the anthropocentric remedies that it attempts on the way) at most will delay rather than avert catastrophe. The real issue is the underlying character of our thinking, and the real catastrophe is that currently we are insufficiently aware of this. And by speaking of the underlying character of our thinking here, I re-emphasize that it is not merely limited aspects of it that need to change. Rather, foregoing argument has revealed the need for a change that occurs, as it were, from the ground up: a reorientation that amounts to a different frame of mind – or way of being in the world – that transcends anthropocentric motives. This means that in educational terms, the focus moves from the epistemological – the generation and transmission of cognitive knowledge (that by itself could simply continue to feed the metaphysics of mastery) – to the ontological: an exploration of different ways of being towards the world and what we are becoming both individually and culturally.[1]

In very general terms, this leads to a twofold approach to environmental education:

1 A short-term pragmatic agenda of damage limitation that focuses on the cautious but imaginative use of science and technology to monitor and help ameliorate undesirable outcomes of the impact of human behaviour on nature. This agenda is now being addressed to some degree;
2 A longer term agenda of developing a sense of a right relationship with nature as the self-arising – this growing sense gradually, but increasingly, informing and orientating the more immediate agenda listed above. This, the most important agenda, is constantly in danger of being peripheralized and subverted by the dominant metaphysics of our time that can only permit it as a façade, a public relations exercise that effaces what is really underway.

And returning in a summary way to the matter of the place of science and how it is revealed from the perspective that I have developed, clearly the issues are complex and tensioned and in part will depend upon what kind of science is being considered. For example, natural history that aspires to carefully observe and record without invasive interventions is clearly of a different kind to experimental science. Indeed, on occasion proponents of the latter have referred disparagingly to the former as mere "stamp collecting" (Johnson, 2008) – the implication being that "real" science is more thrusting, it interrogates and theorizes – perhaps harking back to Francis Bacon, it is not content merely to observe nature, but rather it is prepared to force her secrets from her by whatever means. Indeed, even natural history, when it becomes preoccupied with classification, can be drawn into an emphasis on categorizing and ordering that distracts attention from the native occurring of things. For reasons of this kind, there remains an important sense in which we might learn at least as much, and perhaps more, about nature from the work of Gerrard Manley Hopkins as from that of Isaac Newton or Albert Einstein. If this is true, clearly there are major implications for the conduct of environmental education and what might be involved in ecologizing education.

Yet at the same time – just as with poetry – it is certainly the case that science, particularly in its less aggressive forms, can at some level reflect and initiate a sense of deep wonder and respect for nature. In saying this, I do not seek to deny that there is an important place for more abstract scientific (often computer generated) models, that, for example, reveal salient aspects of what might be happening to the climate or the oceans on a global scale. My point is one of relative emphasis and seeking a balance between alternative approaches – including perhaps seeking opportunities for ameliorating the effects on our thinking of its colonization by the essentially mechanizing procedures and structures that some kinds of science and much *scientism* employ or promote. Indeed, here an important question is provoked from the perspective of ecologizing education: is it possible to create an intelligible space for a kind of science that is somehow poetically infused or energized in the sense that to a significant degree it is informed, both in its procedures and the topics selected for investigation, by poetic intuitions that express some of the qualities of nature in its occurring that have been highlighted in previous discussion? Here I include, for example, intimations of nature's transcendence, autonomy, inherent mystery, integrity and internal norms (including the energizing elemental powers that infuse it), its intrinsic value, and normativity – and very importantly, its "voice". How might a reinvigorated sensitivity to these things be reflected in scientific thinking and activity? What difference could this make to scientific ideas of objectivity and experimentation? Some possibilities in this regard will be explored presently, but overall this remains an important issue for environmental philosophy and the philosophy of education to address.

But whatever more precise answers to these questions turn out to be, in general terms of an aspiration to ecologize education, there will be a need to re-emphasize those aspects of the curriculum that celebrate more receptive and less aggressive relationships with the natural world. In this way, amongst other things, the possibilities of nature's agency in the educational enterprise – indeed, of nature becoming a *teacher* – are enhanced, and an understanding of what would be involved in living in a way that better respects nature's internal norms can materialize. These could constitute important elements of what would be involved in the idea of our "being-in-nature" that was held to lie at the heart of the ecologization of education in Chapter 1.

This being the case, in very general terms, an ecologized education will seek to disrupt the metaphysics of mastery and the blinkers of scientism by celebrating experiences that counter the emphasis that these place on abstraction, prespecification, systematization, and reification. If previous argument in this book is right, it would do this by foregrounding the intimate, the embodied, and the emplaced. Undoubtedly, this will involve a renewed emphasis on the value of time spent in direct experience of the natural world, but not exclusively so. In the time of the holding sway of the metaphysics of mastery, a key parallel educational consideration will be the preparation of the mind–body–heart to be properly open to the occurring of nature in such experiences. For reasons previously indicated, for us moderns, this requires a substantial modification of our intentional outlook in order to receive what is on offer – a suspension of the mastery motive and the adoption of an attitude that is neither an indifference nor a possessive desiring, but rather a *dialogical openness* that incorporates a sense of the well-being of things themselves. The native occurring of things in nature cannot reveal itself to the eye that primarily seeks to organize, to manage, and to manipulate. This is nicely expressed in an example given by the anthropologist Henry Sharp. Drawing a contrast with the other-orientated etiquette of the Chipeweyan Indian, he described "White Canada's" interaction with the non-human world in the following terms:

> White Canada does not come silently and openly into the bush in search of understanding or communion, it sojourns briefly in the full glory of its colonial power to exploit and regulate all animate being. . . . It comes asserting a clashing causal certainty in the fundamentalist exercise of the power of its belief. It talks too loudly, its posture is wrong, its movement harsh and graceless; it does not know what to see and it hears nothing. Its presence brings a stunning confusion heard deafeningly in a growing circle of silence created by a confused and disordered animate universe.
>
> *(Sharp, 1988, pp. 144–145)*

In order to restore to consciousness the fullness of its ecstatic potentiality to perceive and participate in the natural reality that lies before it – giving sight to its looking, hearing to its listening, feeling to its touch – it may well be that not only does it need to engage in the difficult conversation that explores and acknowledges the manner and extent of its current complicity in a frame of mind or way of being that occludes these sensitivities, but also it might benefit from engagement with cultural resources that can provide reorientating narratives and experiences that provoke an attitude of seeing things afresh. As has been intimated in earlier chapters, art and literature might be pre-eminent here. I now turn to some discussion of their potential in this regard.

Art

In a time when the metaphysics mastery holds sway, a key issue is the way in which nature is revealed. It has been argued that in order for things in nature to occur as themselves the attitude of anthropocentrism that ultimately lets them appear only as a resource needs to be disrupted: we need to develop an open attentive receptiveness that allows us to encounter them afresh, freed from masterful instrumentalism.[2] Experience suggests that providing the space for such a fresh encounter is precisely what art can do and therefore it holds promise as an avenue for revealing nature and therefore as a contributor to the ecologizing of education. However, to fulfil this promise, art in education needs to be of a particular character. Its focus needs to be on the work of art conceived not as some kind of representation or statement that necessarily involves an objectification of its subject, but rather, as Heidegger (1975) has argued, as creating a space in which the world in which the thing has its being is invoked, because it is only here that it can appear as itself. To the extent that it is removed from this world, the thing is shorn of relationships that are key to its standing forth in its original significance, such that perhaps now it can only be present as a curio. This view is entirely consistent with the phenomenological account that I have given of the emplaced native occurring of things. Here emphasis was placed on the mutual sustaining involved in the occurring of all the inhabitants of a place, and through which what we might now term their "world" is upheld.

It follows from this that the great value of the work of art from an ecological point of view is that it can – insofar as it works, and through our participation in it – bring us to an awareness of things that is not dominated by a masterful ordering through invoking an experience in which all such ordering is felt to be out of place. In this kind of participation in the work, the thing is not experienced as being represented or defined, and thus is not set at a distance convenient for impersonal rational inspection of it. Rather, it is just there, actively and engagingly, its living presence, as it were, reaching out towards us, penetrating us, such that we can respond to it – receive it and sustain it – as it is, in its own world. Viewed in this way, contemplation of an artwork would have a different emphasis from what is often the case in educational contexts. For example, discussion of techniques employed by the artist and speculation regarding his or her intentions and feelings in producing the artwork – all such "connoisseurship" – would be very much of secondary importance (if not downright distracting), the crucial thing being to preserve that which is actually occurring in the work: the offer of a mutually ecstatic relationship with the thing portrayed. Talk, if talk is needed, would focus on the world of existence that the artwork evokes and the living character of things as they are revealed in that world, rather than the techniques by means of which this revelation might have been achieved. Discussion of the work would therefore arise from the renewed awareness of this world and from within it, and would be of a poetic rather than a disengaged rational character. This presents a stark contrast with the notion of a work of art as essentially some sort of statement and indeed shows this up as a rationalization of art that hides and subverts its vital potential.

In this respect, it has been argued that it is possible that installing a natural place into the space of an art exhibition can help to disrupt routine modalities of perception that are imbued with instrumental rationality. Gudbjorg Johannesdottir and Sigridur Thorgeirsdottir (2015) discuss this in relation to the example of Olafur Eliasson's work *Riverbed*, drawing upon Ronald Hepburn's (2004) notion of "the metaphysical imagination". This latter is conceived as a part of aesthetic experience that senses how things "really are" by sensing the greater (not always articulable) forces that are at work in their being and situating them in the greater wholes to which they belong, hence underscoring the relationality and interconnectedness of all things.

Johannesdottir and Thorgeirsdottir describe how Eliasson's work – a landscape of rocks and gravel meant to emulate a riverbed – filled an entire wing of the

Louisiana Museum of Modern Art in Denmark in 2014. The artist's response to the question of why it was better to go into the museum rather than being outside where nature "really" is was that "Outside we take nature for granted. And when we take it for granted it is like we don't see it". By bringing the environment into the museum as a place where the viewer is used to entering the aesthetic mode of experience in which her senses are open to receive rather than impose meaning, the environment is no longer simply taken for granted. Rather, it is suggested that through the embodied aesthetic experience of walking the riverbed in this space – treading the gravel and touching the rocks – attention can be drawn, for example, to the sheer massiveness present in nature and this can provoke reflection and the awakening of the metaphysical imagination in which we could see

> one particular riverbed "as" the slow movement of the glacier through the ages, or one particular hot spring "as" the volcanic activity of the earth; these forces can even be seen as the creative force of the earth as a whole.

Here it is envisaged that the visitor connects the landscape to the rest of its world and experiences herself within it. Stimulated to pay more attention to our sensory perception evokes an embodied sense of place – and world – in which things are experienced how they really are from within the larger perspective of their own world. Clearly, to the extent that this is, indeed, the kind of encounter that is invoked, such experience of art could have an important role to play in restoring environmental consciousness to its fuller ecstatic potential. Perhaps here, too, arises an insight that contributes to the suggestion that we need to seek ways of poeticizing science. In the light of the power of art to invoke experiences of the kind described previously, Johannesdottir and Thorgeirsdottir criticize traditional Western views of a rationally autonomous self, or one centred around economic self-interest, as assuming that the ethical self can stand independent of its biological and social environment:

> Being human is not characterized first and foremost by having a detached rational mind, but rather an engaged relational body that affects our rational approach to reality. Therefore our emotional response, our embodied aesthetic perception, can and does inform our rational, scientific view of the world. We have been too narrow in selecting what emotions are allowed to inform rational worldviews, and emotions evoked by experiencing the spatial and temporal dimensions of the environment need to be better accounted for in environmental decision making.

If art has the power to make our sometimes vague emotional responses to nature more explicit and to render sensory conceptions such as "massiveness" tactile, it can help us to better understand what they are – what they *mean* in the context of direct bodily experience. Concretized in this way, such knowledge both enhances our awareness and our grasp of what is really going on in nature and consolidates our sense of ourselves as embodied, earthly beings. Hence, such embodied knowledge and deepened sense of emplacement on the Earth increases our sense of connectedness with the non-human world. This, in turn, is a precondition of a fuller sense of responsibility towards it and commitment to take seriously the environmental issues that confront us – as expressed, for example, both by nature's voice and by what, say, scientific models (now grasped in a more embodied way) suggest. Furthermore, the felt sense of the body in natural environments and in artistic approaches to it enables us to sense the ways in which our cognition is fundamentally embodied. Indeed, just by being in the environment and being affected by the powers, forces, and events that are constantly occurring there, the body has a primordial knowledge of the environment that can form a foundation for interpretation and concrete sense-making in cognition by assimilating more abstract perceptions. An ecologized education would see it as important that this pre-predicative understanding is not occluded by too exclusive a preoccupation with celebrating the cerebral.

With regard to the centrality of embodied perception and knowledge, the same authors also consider some possibilities of music, through its physicality, to articulate and make tangible natural rhythms. They suggest:

> Rhythms of music that correspond with rhythms of nature are one way of evoking and stimulating a felt sense of how the inner rhythmical sense matches rhythms out there, in nature.

This invites a consideration of the possibility that, in more general terms, participation in musical experience might help to refine the ear in a way that can re-sensitize us to the subtleties of the acoustic landscape of nature.[3]

Corporeality, time, and place in education

This foregrounding of the significance of embodied perception for environmental education has implications that extend well beyond what the arts might have to offer. Clearly, it lies at the heart of claims made on behalf of immersive experiences in nature. The problem here is that while such experiences might on occasion be sufficiently powerful to command a highly receptive response, given the power of the metaphysics of mastery and the array of stereotypes that it insinuates into everyday consciousness, there must be a substantial likelihood of pupils' encounters being structured by preconceptions that occlude such a response. In addition, given the previously discussed influence of the metaphysics of mastery and scientism on educational thinking and practice itself, there must be a substantial likelihood of experiences organized by educational institutions being set up in ways that also work against this. Indeed, the conceptions of time and space that inform much educational practice at a fundamental level are at odds with those conceptions foregrounded by participation in the native occurring of things in nature.

For example, and taking the former, in mainstream educational contexts, typically time for different areas of study and the activities to be engaged in is carefully (often rigidly) structured and measured out. Linear segmented clock-time rules. Discussion undertaken in Chapter 4 made clear that this is precisely *not* the kind of time in which nature participates and that the masterful projection of clock-time upon it occludes our full participation in its occurring such that we are left with a knowledge only of reified objects rather than things themselves. Similarly, the space within which education is conducted frequently is far removed (both physically and psychologically) from the unique and vibrant places in which nature as a dimension of experience can be keenly felt. Undoubtedly, these fundamental oppositions are reflected more broadly in the underlying motives and conventional views on "effective" pedagogic strategies that operate within much mainstream education and that are likely to be in conflict with what is required for the growth of environmental consciousness. The prevalence of economistic motives such as preparation for "the world of work" in consumer-orientated societies and the pre-specification and modularization that frequently follows from this are cases in point. As David Orr (1994) has put it, with regard to conventional Western education that increasingly emphasizes producing students who are effective operators in a global market economy premised on perpetual material growth, in many ways environmental concern raises not simply problems *within* education, but the problem *of* education.

With this in mind, and returning to the matters of time and space identified earlier, it will be helpful to consider a number of ideas that have emerged under the rubrics of "slow" and "wild" pedagogies. Phillip Payne and Brian Wattchow (2009) have characterized the former in the following way:

> A slow pedagogy, or ecopedagogy, allows us to pause or dwell in spaces for more than a fleeting moment and, therefore, encourages us to attach and receive meaning from that place.

In responding to our current ecologically problematic human condition they go on to say:

> There needs to be a shift in emphasis from focusing primarily on the "learning mind" to re-engaging the active, perceiving, and sensuous corporeality of the body with other bodies (human and more-than-human) in making meaning in, about, and for the various environments and places in which those bodies interact and relate to nature.

Here, *inter alia*, the primitiveness of the body and its existence in biological, circadian, and cosmological times will need to be respected. They contrast this corporeal orientation with the "fast, take-away, virtual, globalized, download/uptake versions of electronic pedagogy" that, for reasons of convenience and economy, are in danger of colonizing environmental education. This surrogate for proper – that is "slow" – immersive experiences in nature replaces the latter with "disembodiment, displacement, disembedding, and decontextualizing of varied face-to-face interactions and relations with others, including 'nature'". Clearly, here again, echoing the discussion of the use of digitalized experience in education in the previous chapter, we would have a pretty complete denial of the emplaced intentionality of environmental consciousness, its ecstatic relationships becoming uprooted and confused in a mire of abstraction.

In their discussion of an undergraduate course designed to invite corporeal engagements with natural places through slow pedagogy, Payne and Wattchow make it clear that simply providing opportunities for students constructively to linger – indeed, dwell – in a place requires structured elements that encourage them to loosen and question the grip of historical presuppositions, assumptions, memories, and attitudes. Students need to understand more fully, and in some cases modify or transcend, what they bring with them to their experiences of nature. The subsequent experiential recycling of historical and new bodily experience and remembered and imagined ideas can acquire a renewed intensity in places where elemental nature is to the fore, and the "embodied dissonances" that can occur can amount to "ontologically reassembling the self". Interestingly – and relating closely to previous discussion of nature's normativity – one of the key questions that students were asked to consider during the direct experiential parts of the course was: "What will nature permit us to do here?" This suggests attention to some sensing of nature's normativity by the students. Throughout, the authors observe that silent musing rather than talk can lie at the heart of the growing understanding and becoming of the embodied self, and note the danger in experiential education of the teacher-controlled "debrief" to colonize the experience of participants. Referencing the account of environmental consciousness that I have given, what ultimately must lie at the heart of experiential education is freedom of response and articulation so that the integrity of the experience is preserved. Clearly, this is antipathetic to pre-specification of learning and any kind of coaching towards "right" or accepted answers.

It seems to me that the spirit of this stance is nicely illustrated by the philosophy of "wild" pedagogy, to which one imagines that Payne and Wattchow would subscribe. Marcus Morse, Jickling, and Quay (2018) characterize wild pedagogy in the following terms:

> Wild pedagogies are about rethinking our relationships within the world and represent a desire to let go of an overabundant sense of control, to invite the places we visit to become an integral part of our educational work and to respond to provocations in spontaneous, and at times unforeseen, ways. . . . In doing so we situate some of these underpinning ideas within touchstones – intended as provocations and reminders of what we are trying to achieve. The touchstones described include: agency and the role of nature as co-teacher; wildness and challenging ideas of control; locating the

> wild; complexity, the unknown, and spontaneity; time and practice; and cultural change. These touchstones are drawn from experiments in practice and attempt to bring the more-than-human world actively into educational conversations.

At the heart of this heuristic approach is the aspiration to experience nature afresh where the "wild" in "wild pedagogies" refers not only to an aspiration to go with the flow of what is experienced unfettered by overweening conventional preconceptions that hinder the arising of imaginative possibilities for humans to dwell in places, but also a conception of nature derived from Old English etymology where wilderness is understood as "self-willed land". Nature as the "self-willed" seems to resonate strongly with the idea of nature as the self-arising that informs the current work and to embrace the themes of nature as having its own integrity, internal norms of development, and, indeed, normativity, that have been the subject of ongoing discussion.

A nice illustration of attempting to provoke a refreshed experience of nature is given by Morse, Jickling, and Morse (2018) in an account of using pinhole photography on the Franklin River, Tasmania. The use of this (technologically) unsophisticated photography in which there is no lens or viewfinder (and therefore reduced power to frame and pre-specify what will appear) is regarded as an experiment in paying attention that invited participants to slow down and to attend and listen to the surrounding environs and to encourage new ways of perceiving and relating to a place. As the authors put it: it is a "deliberate pedagogical approach that sets out to provide a fertile field for purposeful experiences without necessarily controlling the outcome in advance: a wild pedagogy". Pedagogic strategies of this kind can create an educational situation in which nature's agency can participate more fully and in this sense students' experience is ecologized. Clearly, from the perspective developed in this book, "slow" and "wild" pedagogies as heuristic approaches have much to recommend them, but in broad practical educational terms a caveat has to be entered: along with other approaches that centre on immersive experiences in nature, how is this experience to be made available to all who might benefit from it? This is an issue to which I will return.

For the present, and with regard to issues of time and space/place, the account as it now stands has made two important considerations clear. First, it should not be assumed that simply arranging for pupils to be placed in natural places will provoke the kinds of encounter with nature that will develop environmental consciousness. Second, some of the concerns that arise in response to this raise important issues not only for learning in the context of environmental education, but for learning in education more generally. They suggest a kind of learning that has a very different orientation to that which is dominant in many educational institutions. The first of these considerations provides an entrée into a more explicit treatment of a topic that has surfaced at various points in previous chapters: the place of poetry (but also literature more generally) in developing environmental consciousness. I will address this in the section that follows. The second set of issues concerning implications for education and the philosophy of education in general will be taken up in the final sections of this chapter.

Literature

By way of preliminary to examining the potential of literature to contribute to the development of environmental consciousness – its extension and refinement – it will be helpful to return to a general issue concerning the mediation of nature. Throughout the arguments presented in this volume, some readers might feel that they are able to detect a tension – sometimes explicit, but often implicit – between ideas of being able to experience "things themselves" in nature and the mediating power of concepts and language that is inherent at some level in all such experience. Indeed, there are those who have argued that what we perceive (including our perception of ourselves) is the product of the cultural narratives or "stories" to which we have access and bring to the experience. Following this line of thought, M. J. Barret (2006), for example, speaks of these stories as producing both our experiences and our sense of ourselves and what we can become. She goes on to conclude that "an emphasis on interrogating discursive production of experience and subjects is very much about becoming, and in particular, what environmental education has, and might, become".

Certainly, the position that I hold has a degree of sympathy with this view. However, it would caution against this attention to discursive production becoming the dominant focus of environmental education such as to demote the significance of a direct attendance upon nature. Indeed, with regard to one of the central themes of this book, to the extent that our experiences of nature are, indeed, discursively produced, is not (as Barrett implies) the idea of experiencing a transcendent nature an illusion? It seems to me that there is a danger here of conflating two related but distinguishable things: (a) the experiences of nature that we happen to have (for whatever reasons), and (b) the possibility of experiences that can indeed be considered in a significant sense to be of things themselves in nature. For example, it has already been granted that there are powerful narratives that heavily condition our experiences of nature: notably how under the sway of the metaphysics of mastery essentially nature is experienced – as it were, "storied" – as a resource. The aspiration of much subsequent discussion has been to demonstrate that in many critical ways, this is a cabined and inadequate experience of nature and that a more openly receptive consciousness is possible – one that can become acutely aware of this effect. It is one thing to claim that the stories to which we have access can be an important factor in what we experience, it is quite another thing to claim that they alone determine experience.

It is important to note that saying this does not deny that our human form of sensibility plays a part in such experience – far from it – but it has asserted the agency of something that lies beyond this sensibility (and indeed, as being entailed by it) as also having a role, such that it is defensible to hold that there can be a primordial experience of things themselves as they arise for human beings, and that this experience is given to us as we pay proper attention to the native occurring of things. But, again, this is not to say that stories might not have an important role to play in ecologizing education. Given the need to disturb current powerful ways of viewing nature that curtail the ecstatic potential of consciousness with regard to it, and in line with considerations that were raised previously concerning the potential of art to play a part here, it is important to consider (albeit briefly) the contribution that literature might make to provoking the imagination in ways that might transform experience of nature.

Reference has already been made in previous chapters to ways in which great poetry can (re)sensitize us to nature as a vibrant dimension of experience and to the native occurring of things in nature. Wordsworth and Hopkins are pre-eminent here and the "sprung rhythm" of the latter that so ably communicates something of the "selving" of things in nature and their individual "inscapes" is particularly salient and was discussed in Chapter 3. In his discussion of mythopoeic literature, Matthew Farrelly (2019), drawing on the work of Martha Nussbaum, shows how imaginative stories can transform how we engage with the world and that, drawing on Joseph Campbell, myth, understood as "the secret opening through which the inexhaustible energies of the cosmos pour into human cultural manifestation", gets to the bedrock of reality and meaning. Insofar as this is true, through his art the poet can perceive and articulate aspects of this bedrock such that the archetypes that secretly inform our intuitive understanding at a deep level can arise from the undergrowth of ordinary fact and become manifest. These supply, as it were, touchstones of human sensibility that send thought on its way by suggestion rather than explanation.

Farrelly shows how in this regard the poetry of Hopkins, through evoking the individual "personality" of things in relationship, is suggestive both of what they are in their individual essence and hence what they call for (what I have previously termed their normativity) and of how one ought to respond to them so perceived, including when somehow violence has been done to them. Here, in foregrounding this deep revelatory quality of poetry, there are strong resonances with both Heidegger's (1975, p. 215)) discussion of the idea that "poetry is what really lets us dwell", and with Magrini's views on nature's call to us and its ethical challenge (previously discussed). Overall, there would seem to be a strong case for claiming that engagement with poetry as part of aesthetic education, through evoking an intuitive imagination, can contribute to an enriched perception of the natural world that constitutes its authentic re-enchantment.[4] By the same token, this will fulfil some key positive possibilities of ecstatic consciousness by reconnecting it with a felt intuitive reality: a wellspring of significance that lies at the heart of human dwelling.

The place of wonder in education

Perhaps one of the most common phrases to be encountered in the literature of nature appreciation and education is that of experiences of "awe and wonder".

And not without good reason, for this seems to be a wholly appropriate response to the majesty and beauty that nature can present. Experience of this kind can be a powerful motive for getting out into the natural world and re-establishing contact with it – something that many see to be a prerequisite of gaining environmental understanding. Indeed, Haydn Washington (2018) has argued that the experience of wonder towards nature is a key factor in supporting the degree of belief and commitment necessary to take the difficult decisions that we now face in addressing our current environmental crisis. Here he follows Anders Schinkel (2017) in distinguishing between wonder as in the sense of being curious and "deep wonder" as feelings of astonishment, amazement, enchantment, and love that can be experienced when in the presence of something magnificent or laden with significance whose existence is gripping and mysterious, sometimes to the point of appearing miraculous.

Clearly, experiences of this kind are ones to be enjoyed and savoured, and in themselves instantiate a deep and sometimes unexpected sense of connectedness with nature in its self-arising otherness. Here we have something that is the polar opposite of what has been termed "the extinction of experience" of wild and elemental nature that pervades everyday life in late-modern society, and that has the potential to counteract it – to rejuvenate a sense of a larger world beyond, of which we are a small and often relatively small and powerless part. Furthermore, it could be said that this positioning of the individual in respect of the world is apposite not just in the context of environmental education, but also in the context of learning more generally and in this sense represents another avenue for the ecologization of education. The humbled and highly receptive posture that it depicts is perhaps central to the genuine seeker of truth and certainly resonates closely with the ecstatic nature of consciousness elaborated in Chapter 4.

But at this point it is necessary to enter an important caveat. While clearly giving the impression of consciousness being in the grip of something other than itself and in this sense highly ecstatic, experiences of awe and wonder are not metaphysically innocent. Like all other experiences, they occur in the context of an extant metaphysical climate. Therefore, an educational emphasis on developing a sense of wonder needs to be alert to the cultural conditioning and anthropocentrism that can structure it. This is to say that under the auspices of developing a sense of wonder, nature can be set up as a provider of particular culturally conditioned affective experiences such as the aesthetic pleasure associated with the perception of beauty or awe in the face of the sheer power or majesty of elemental nature. Set up in this way, cultivating a sense of wonder can lead to both a stereotyping of the kinds of experiences to be had in nature and a lack of attention to what nature itself might communicate if the foci of our attention were less preformed. There is a danger that less prominent, sensational, immediately impressive aspects of nature will be overlooked, and that the wonder to be experienced, for example, by carefully and patiently attending to small-scale phenomena over a period of time will be occluded. The flow and many countenances of coming into being and passing out of being – revealing and withholding – that occurs in, say, a patch of vegetation underfoot can be as much source of wonder, and can communicate as much of nature, as the observation of a splendid sunset or the courses of the planets. While not wishing to undervalue the undoubted good that, say, some TV documentaries can do in bringing to the attention of a wider audience vivid examples of the integrity, beauty, and majesty of nature – and the destruction being wrought upon it by human behaviour – there can be a tendency to focus on the magnificent – perhaps the sensational – that distorts by exclusion a proper picture of the extent of what is really afoot. The central point here is that we need to cultivate a genuine receptivity and openness to the being of nature as it occurs from out of itself rather than to view it exclusively through a lens of pre-formed expectations or culturally established proclivities and the seeking of aesthetic thrills.

It has been argued that drawing a distinction between the ideas of *haecceity* and "suchness" can make an important contribution to understanding this. In contrast to my argument in Chapter 3, Michael Chang (2020), who is very sensitive to the kinds of reservations that I have voiced earlier, argues that tacitly *haecceity* refers to the definitive characteristics of a thing that involves a rational attempt to uncover the mystery of nature by coaxing a definition from "mute material". He contrasts this with "suchness" that he describes as referring to the expression of things themselves without human affective projection or addition. For him, when we experience suchness, we perceive the world just as it is: "a reality entirely sufficient and vibrant in itself", free of the embellishments of a human subject. It seems to me that attractive as this idea is, it is also very problematic. While offered as a statement of fact, it is really a *claim* and, if interpreted in a radical way, a very contentious one at that. By "radical way" here, I refer to the idea that somehow in particular experiences we can perceive things simply as they are in themselves in the sense of their being perfectly autonomous, the manner of their standing forth owing nothing whatever to the consciousness that beholds them. Here, experience of their appearing would transcend or undercut our form of sensibility with its sensory and conceptual structures, linguistic articulations, and its proclivities and other cultural determinations.

Now the question that arises is: is this intelligible? On the arguments developed throughout this book, which demonstrate that interpretation is present in, and essential to, *all* perception and understanding, it is not intelligible. Everything is something. And what this something is, ineluctably is partly conditioned by the place in which it shows up: a perceiving consciousness. If we are not positing some mysterious realm of things in themselves that are the source of all perception – and that by definition cannot be perceived – and hence could not lie at the heart of ideas of suchness, where sheer appearing is all – then we must accept a less radical – but not any the less significant – notion of suchness. This less radical interpretation of suchness would accept that the play of being always occurs within and against a form of sensibility. On this account, the force of the idea of suchness is not to conjure some esoteric experience that denies this, but to behove openness. It exhorts us to eschew fixed preconceptions and ways of perceiving that reify and deaden things that are capable of standing forth as spontaneous and vibrant in ways previously described. On this interpretation, suchness leads not to a denial of a conditioning form of sensibility, but to its "purification" of all that wilfully effaces and occludes salient aspects of what lies before it. In practical terms, this would support not only practices of free play in nature in the sense of encouraging spontaneous responses to the play of being to be experienced in natural places, but also invitations to participate in ways that disrupt routine responses and that lead to questioning of currently dominant attitudes and perceptions of the kind previously noted.

The idea of human stewardship

One very popular and indeed evergreen (dating back to *Genesis*) notion that is taken to characterize how humanity should relate to the rest of nature is that of stewardship. In the respect that it embodies a caring for nature this notion might claim to have much to recommend it – indeed, it might be taken to express very felicitously the kind of responsibility and commitment that humanity should demonstrate towards the natural world. However, in the light of arguments developed in this book, this seemingly benign notion contains a flaw that infects so much modernist humanism and if not identified and expelled would prove fatal to achieving the kind of relationship that is required. Although to a degree veiled, it is resonant with the metaphysics of mastery. Not only does it set up humankind as being superior to the rest of nature, it sets up nature as in need of being managed by us for its own good – and on an increasingly grandiose scale. Hence, sometimes we hear of ambitions to manage forests, wild fish stocks – even the oceans. Apart from the previously discussed question of the inadequacy of our knowledge base and predictive powers in the face of the sheer spatial and temporal extent over which natural systems operate, there is a tacit reversal of what needs to be managed that amounts to a normalization of anthropocentric interference in nature. When we speak of, say, the need to manage wild fish stocks, we should be talking of managing human behaviour towards a fish population; left alone, this population is quite capable of "managing" itself.

One aspect of the extent of the humanistic hubris involved here is emphasized in the following argument made by Stephen Gould (1990). He observes how frequently two linked arguments are promoted as a basis for an environmental ethic. The first is that we live on a fragile planet now subject to permanent derailment and disruption by human intervention. The second is that humans must learn to act as stewards for this threatened world. He raises the thought that such views, however well-intentioned, are rooted in the old sin of pride and exaggerated self-importance and goes on to make the following point:

> We are one among millions of species, stewards of nothing. By what argument could we, arising just a geological microsecond ago, become responsible

> for the affairs of a world 4.5 billion years old, teeming with life that has been evolving and diversifying for at least three-quarters of that immense span? . . . This assertion of ultimate impotence could be countered if we, despite our late arrival, now held power over the planet's future. But we don't. . . . We are virtually powerless over the earth at our planet's own geological time scale. All the megatonnage in our nuclear arsenals yield but one ten-thousandth the power of the asteroid that might have triggered the Cretaceous mass extinction.

This echoes some aspects of the discussion of the idea of an Anthropocene epoch broached in Chapter 1. Essentially, its focus is upon a perceived lack of human capacity to control what is afoot on the scale relevant to global stewardship. But in the context of developing an environmental ethic, a further dimension is surely apposite: the issue of whether we should seek to extend our power and aspire to this role. It is one thing to do our best to repair the damage that we have done to natural systems as and where we are able; it is quite another to seek the general mastery of things that is implied by stewardship. This, of course, brings us back to the character of our underlying relationship with nature: should we set ourselves up as managers and if so, managers of what precisely – and in whose interests? Previous discussion suggests that essentially the "interests" of the natural world will be best served if we eschew the idea of stewardship of a nature that has its own complex and intricate facilitating and feedback processes and focus on managing our own behaviour towards it so as to minimize interference. This reference to the character of our role in relation to the natural world leads to the more general and complex issue of what should be our underlying moral stance.

Moral and social education

Previous discussion suggests that an ecological focus will have extensive implications for this area. If nature, or some aspects of it, possess intrinsic value and therefore moral standing, clearly what we regard as the moral community is considerably extended compared with those conventional views of morality that are founded on the idea of morality as a contract or other relationship that exists exclusively between rational moral agents (as with, say, Kantian ethics). This was illustrated in the discussion in Chapter 5 of the idea of ecological justice in which it was made clear that moral choice and action must take into account nature's needs, and in a way that if these latter are not conceived as being on exactly the same footing as human needs, that does not make them entirely subservient to human needs. *Inter alia*, this means that the moral standing of nature is not simply dependent on its potential to serve what are taken to be human needs in either the short or long term, and therefore we are obliged to be attentive to nature's own normativity. In broad terms, this foregrounds the idea of nature's voice and the character of the kind of attentiveness necessary to "hear" it. It was argued that this consideration puts an intuitive receptive-responsive sensitivity to the other at the heart of morality. In this way, in being open to the native occurring of things in nature, we can receive intimations of what would be fitting responses. As James Magrini argued, we need to be receptive to nature's call and to allow it to condition, but not determine, our ethical response.

In broad terms, this position raises two sets of issues for moral education that derive directly from the central themes of this book. The first set of issues concerns the establishment of a fuller understanding of – and hence proper relationship with – the natural world. The second set is concerned with how this renewed orientation might transform moral education as a whole, including now human–human morality. Throughout this account the transgressive distorting influence of the metaphysics of mastery on our way of being in the world has been a pivotal theme and its relevance to both these sets of issues is clear. An overweening (if nonetheless frequently tacit) desire to master what lies before us that transforms everything into a resource can hardly do otherwise. Throughout there has been frequent mention of the need for education to disrupt this metaphysics – in general, by making it visible and questioning its legitimacy. It might be helpful here to provide a more detailed account of the sort of thing that this could involve by explicating the first set of issues in terms of a specific illustration: the need for a "decolonization" of nature.

Drawing on the work of Albert Memmi, Blenkinsop *et al.* (2017) have developed the perspective that the relationship between modern humans and the natural world is one of colonizer and colonized. There are many strands to this account, but a key one is the idea of developing sustained practices of listening to the colonized other and ensuring that any theory or action designed to "solve" environmental issues follows from this engagement rather than being superimposed upon it. They note how "hierarchizing and amalgamating myriad individuals into a single grouping is a classic colonial strategy" and that this, coupled with a "distrust in the natural world's capacity to "self-govern", pervades the ways in which humans attend to "environmental issues". This is reflected in conventional Western education which they summarize in the following terms:

> Schools, for their part, manufacture the Colonizer and the Colonized vis-à-vis the layout of the campuses, the structure of buildings, the institutional norms, and the school curricula. In each of these, there is the implicit understanding that the "natural world" is a distraction to learning, that humans can and are destined to shape the world solely for their own needs, and that knowledge is structured and located within either the teacher or the rows of books in the library. Other beings are systemically silenced and cultural practices and ontologies that suggest otherwise are presented as archaic, sentimental, or uncivilized.

As a key part of their education, they note how "From an early age, colonizers are taught to not-know that they are deeply interwoven in a fabric of interspecies kinship". This is a form of "bad faith" that enables them to benefit from the violence done to the non-human world without feeling morally culpable. But in Memmi's words, the destruction of the colonized also "rots" the colonizer.

It seems to me that there are numerous synergies between this view and the account that I present in this book, and that, indeed, it nicely expresses some of its central themes. In particular, it emphasizes ways in which the issue is not one simply of the formal curriculum, but the underlying culture of the school. In other work (Bonnett, 2010), I have noted how this culture conditions both the ways in which things are presented (including the formal curriculum) and the ways in which they are taken. Also, I noted there how a culture that expresses the metaphysics of mastery leads us to expend ourselves in ultimate futility: caught up in essentially circular efforts to manipulate, control, and consume, we dissipate our capacity for genuine engagement with the world – a mutually affirmative creativity that through being open to otherness, what is gifted, can refresh and inspire. In the absence of this, we are doomed to a form of mental constipation in which, indeed, as Wordsworth has it, "We lay waste our powers".

However, accepting these considerable synergies with the view of colonization of nature described, it is worth noting a potential point of difference: Blenkinsop *et al.* speak of "Nature" being reified as the source of Truth and as a mythological ideal that is enigmatic and transcendent, both dangerous and wondrous. They say that these descriptors "work effectively to maintain a sense of unknowability which perpetuates a seeming ontological gulf between humanity and other beings, in a form determined for the purposes of the colonizer". Depending on precisely how notions of "transcendent", "unknowability", and "ontological gulf" are being interpreted here, there is the potential for a significant tension between this account and my own to arise. Throughout, I have consistently emphasized both the significance and peculiarities of human consciousness (or its equivalent) and the inherent otherness and mysteriousness of nature, and have criticized views that seek to elide all fundamental differences between human and non-human nature – often as either anthropomorphic or reductionist. Hence, when, in this regard, Blenkinsop *et al.* go on to speak of a need to challenge "out-dated descriptions of the 'natural world' (often not even valid according to current biological theory)", I would caution against casting as "out-dated" conceptions that arguably are primordial to our form of sensibility (see, for example, my response to Affifi's use of "enactivist" notions of cognition in Chapter 5). I would also express reservations about the scientism displayed in the assumption that current biological theory should be regarded as arbiter of what counts as natural. Perhaps, the high degree of equality between the human and non-human implied by Blenkinsop *et al.* is a function of their taking as their original point of reference an account of colonization in a human–human context.

Nonetheless, reference to their account prompts some clarification of the human–non-human distinction that informs this work. What I have termed the "great divide" is considerably more modest than traditionally conceived and implicitly expressed in modern culture, but this does not mean that there are not very significant differences between the human and the non-human – as previously discussed, for example, with regard to kind of consciousness and the extent of its temporality, capacity for choice, imagination, moral responsibility – and, yes, rationality.

With regard to the issue of the overall character of moral education, a number of aspects come into view. It seems to me that the renewed receptivity to emplacement, otherness, and non-rational agency that a full engagement with nature foregrounds is highly suggestive for understanding morality in the human–human context. For example, in contrast to one still highly influential characterization of morality as "a set of principles to govern human life in general" (Barrow, 2007, p. 28), it follows from an account of the kind developed previously that fundamentally ethical concern does not arise in some pure form of the kind that can adequately be articulated in sets of universal abstract principles. Rather, it is the case that our antecedent involvement in a place (and therefore the world) conditions all understanding, including the ethical. Here arises a recognition of the extent to which moral life is concerned with such and such an issue in such and such a context, *and some aspects of the participatory nature of that context.* Hence, there is an intimate reciprocity between ethical and environmental concern that fundamentally initiates the character of our caring. Such a position invites exploration of an idea of moral education that premises the idea of moral agency on that of poetic receptiveness. In doing this, it intimates an enlarged sense of moral agency: one that is less exclusively preoccupied with the model of an autonomous rational agent and that seeks to sensitize us to possibilities of an enabling passivity on our part. This would involve a proper recognition of the play and value of non-human agency and the ways in which there is a reciprocal co-responsibility for the character of the places in which we live. Clearly, as previously noted, this represents a radical challenge to the traditional ethical position of conceiving moral obligation within a social contract that can only hold between rational agents.

From the perspective of transcendence, ecstasis as an open ability to stand beyond ourselves, rather than empathy or sympathy, becomes what is most fundamental to being moral: allowing the other to be itself in all its mystery rather than presuming to know in advance – to fully comprehend – its feelings and purposes, and indeed whether or not it has feelings and purposes. In this view, it is not so much that on occasion, rightly, our actions might be informed by a sense of what someone else could be feeling, it is that we need to be alert to the ever-present danger of responses that express our own proclivities and expectations prescribing *their* sense of what they should be feeling. Intimately emplaced life is too subtle and multifaceted for this to express true moral regard. We must acknowledge the essential otherness of the other. This is particularly salient when it comes to the application of notions of social justice. Here a proper regard for the otherness of others – their essential unknowability – leads to an acknowledgement of the centrality in moral culture of personal freedom, and responsibility for the consequences of its exercise.

When allied with previous argument concerning the ontological uprooting and unselving of things when, in thought, they are abstracted from their emplaced existence through being rationally categorized and quantified in preparation for "equitable" treatment or distribution, it can be seen that the standardization involved is out of kilter with the lived reality both of nature and a consciousness that is released towards it.

Pedagogy and the curriculum

In addition to raising issues concerning the potential of different areas of the curriculum to contribute to ecologizing education, a question arises regarding the general organization of a curriculum that promotes the form of learning necessary to reconnect with nature. As has been made abundantly clear – and was rehearsed at the beginning of this chapter – the focus of ontological education is on attitudes and approaches to the world rather than the acquisition of rationally organized knowledge. In addition, the character of "natural" time and "natural" space alluded to in a previous section and discussed in earlier chapters has implications for engaging with things in the receptive-responsive way that previous argument has commended. With regard to the former, the point was made that proper attentiveness to nature occurs not in linear clock time, but in timeless moments and the "spiral" time of cyclical rhythms. Here the learning consciousness needs to be both unfettered in its absorption with things that it beholds and to exhibit patience and a preparedness to revisit phenomena so that it can be attendant on new nuances of what is occurring and become aware of ongoing processes. This suggests that the curriculum itself should have this spiral character. In general terms, this is to say that the organization of time in education should reflect the quality of time that things bring with them rather than that superimposed by a will that is set on meeting pre-specified "educational" objectives within a restricted time frame. Similarly, with regard to space, education needs to respect the essential emplacement of things in nature. This requires that where possible learning should be situated in more natural places where nature's agency can be felt and become part of pedagogy. It also requires sensitivity to the degrading effects of things learnt extracted from their place – the place in which they have their own meaning. Taken together, these considerations indicate an emergent curriculum in which pre-specification is kept to a minimum and discipline boundaries are regarded as highly permeable.

This returns us to the question of the relevance of immersive experiences in nature as an educational strategy in the late-modern period. For example, how are we to provide access to experiences of the native occurring of things in nature to a population increasingly situated in megalapolitan conurbations and perhaps heavily in the thrall of digitalized experience? Regarding the latter, it was argued that, in contrast with authentic experiences of self-arising nature that are spontaneous, embodied, and multi-sensory, here predetermined aspects of reality are called up and manipulated at will and are experienced through sight and sound of a highly restricted kind.

In such circumstances, some might regard a quest to bring these populations into a more intimate engagement with the natural world as simply risible or irredeemably "romantic". Yet if the arguments outlined in this work hold, it must be attempted.

And perhaps the case is not without hope. Clearly, while the kind of experience of elemental nature discussed in the previous chapter of rafting on the Franklin River is hardly likely to be available to all – and certainly not on a regular basis – salutary experiences of the native occurring of things can occur in less wild places that are more widely accessible (as the vignettes of the woodland dell and upland stream presented in Chapter 2 demonstrate). In any case, cities are not without possibilities of such engagements, although in some cases the challenges to finding them are strong. Undoubtedly, the natural dimension can be very tenuous in the life-world of city-dwellers. Indeed, perhaps for some it seems far-fetched to suppose that it could be otherwise. Their environment might be described as one of relentless artificiality, from the tarmac underfoot to the obliteration of the presence of the firmament by light and sound pollution, and to the diminution of the seasons and other natural rhythms – perhaps, even those of night and day – through other human interventions or dominant social habits. In such circumstances, it might appear otiose to think of their life-worlds as in any meaningful sense portals to an appreciation of the native occurring of things in nature. It is not, of course, that nature has no presence at all – rain or snow still fall, the granite kerbstones still have their own solidity, water still trickles, collects in puddles and reflects light in myriad ways, small-scale wildlife might still be found under some stone or growing in some crevice – but, for many, it has no attractiveness and has become invisible. Concern and attention lie elsewhere. There are two things that I would like to say on the back of this kind of example.

First, from an educational perspective, there may well be strong reasons for giving those who lack them opportunities to experience elemental nature in contexts more conducive to attentiveness and where it is more immediately engaging. On occasion money may be better spent on field trips than on, say, digital upgrades.

Second, there is evidence to suggest that even relatively limited nature-rich experiences in childhood can make a lasting positive impact on people's attitudes (see, for example, Palmer, 1999), although clearly this falls short of the kind of re-naturing of human life advocated by Richard Louv. Certainly, the arguments mounted in this book support the aim of enabling a significant element of an ongoing living *in* nature in everyday life such that our sense of connectedness with the elemental and our ability to listen to nature's voice in our decisions is substantiated in this life. Given this, a re-naturing of urban environments wherever possible in tandem with a dilution of the metaphysics of mastery and the development of environmental consciousness are wholly desirable.

Education without a future

However earnestly we might wish it otherwise, a bleak possibility now confronts us. It is entirely possible that anthropogenic impacts on nature either have already set in train processes that will lead to the extinction of humanity in a foreseeable future, or that this is not yet the case, but that it will become so due to lack of adequate and timely action. What difference, if any, does this possibility make to how we should think about education? Does it present issues that the philosophy of education should address? For example, sometimes (some might argue insufficiently) philosophy of education has explored the educational significance of human mortality. Could contemplating the drawing near of the extinction of the human species as a possible prospect yield considerations of educational importance? For example, how might education begin to look if specified longer term future achievements are diluted in our thinking about what constitutes a good life? – Where, perhaps, the strictures of a rampant will are loosened so that attention can return to the quality of the present? Might this provide an opportunity to (re)focus on the character of our current relationships with the human and the non-human, to be more open to the gift of otherness and the call of things from within their own space and time? And in this way, to live more generously?

Ecological selfhood and the hidden centre of authentic education

One of the most fundamental issues raised by the arguments mounted in this book is the nature of authentic selfhood and the character of the relationships in which it participates. Clearly, the account that is given of this will be absolutely central to understanding what it is for someone to engage in education and to become educated. Views of education that have at their heart a notion of self as an essentially self-contained entity that interacts with an external world through its self-deliberated choices and actions and that therefore needs to acquire information that can be fed into its autonomous thought processes to make them better informed will be disconcerted by this view. Even if the former, as it must, concedes that the external world constantly impinges upon the self in ways that it did not choose, such that the self must react and adapt to changing circumstances in order to cope and regain a degree of control, there remains a sense in which the self is essentially conceived as sovereign of its own being. It lives to exercise its will and a key function of education is to enhance this will (for example, by making it more reflective and informed) within a morally accepted framework that is itself the product of this will. Here, then, the idea around which ideas of the self and education rotate is that of degree of autonomy, the development of the rational agent, the one who must *decide* what is to be done. As such, it must be fed consumable parcels of knowledge and skills metered out in a way that reflects what are anticipated to be its future requirements.

While certainly not entirely eschewing a place for the will and its decisions in selfhood, the view that has emerged over the course of this book nuances things rather differently by admitting a greater place for creative passivity in its account. Space is given in which to recognize the agency of what is other that is operative in the places in which we live and in self-formation (which it is argued are in internal relationship). This gives rise to the idea of co-responsibility in the phenomenology of place and of self-formation: the emplaced transcendence of environmental consciousness. If consciousness is indeed ecstatic and therefore inherently environmental in the ways described, when healthy, its life and activity is always to a significant degree elicited through its encounters with what is other rather than, as it were, being exclusively posited from within. While an overall sense of "mineness" attaches to its life and it retains an overall sense of responsibility for its actions, in many respects the self is, indeed, called forth by what lies beyond. Its attention can be claimed by what is other, feelings can be evoked by what is other, actions can be experienced as being sanctioned by what is other. Hence, education for selfhood must embrace a degree of spontaneity sufficient to allow this to occur, to be celebrated, and to initialize further encounter. I have previously claimed that pre-specification is the bane of educational experience because necessarily it obstructs one from being fully *there*, in a place, open to all the nuances of the native occurring of things. Freed ecstatic consciousness, properly emplaced, is the hidden centre of authentic education. And for reasons previously given, this kind of participation reaches its zenith in places that are nature-rich. Hence, these nature-rich experiences lie not only at the heart of an explicitly "ecologized" education, but at the heart of education in general – which is to say that education itself is rooted in this kind of ecologization.

Notes

1 Although, as previously argued, this does not – and could not – entail a complete jettisoning of concepts such as "reality", "truth", "consciousness", and "self" that articulate our form of sensibility at its most fundamental level. Perhaps a distinction can be drawn between this primordial *form* of our sensibility and more specific *modalities* in which it operates – ways in which its potential gets expressed as in some particular set of deep cultural motives (i.e. metaphysics) such as the metaphysics of mastery.
2 I use the term "masterful" here to distinguish this kind of instrumentalism from the kind that receives nature as a gift, as previously discussed in Chapter 3.
3 See, for example, Ostergaard, 2019.
4 See Jeffrey Stickney (forthcoming) for some interesting further exemplification of this.

BIBLIOGRAPHY

Abram, D. (1997) *The Spell of the Sensuous* (New York, Vintage Books).

Affifi, R. (2017) The metabolic core of environmental education, *Studies in Philosophy and Education*, 36, pp. 315–332.

Arendt, H. (1978) *The Life of the Mind* (San Diego, Harvest Books).

Arendt, H. (1998) *The Human Condition*, Second Edition (Chicago, University of Chicago Press).

Ayer, A. J. (1940) *The Foundations of Empirical Knowledge* (London, Palgrave Macmillan).

Barret, M. (2006) Education for the environment: Action competence, becoming, and story, *Environmental Education Research*, 12(3), pp. 503–511.

Barrow, R. (2007) *An Introduction to Moral Philosophy and Moral Education* (Oxford, Routledge).

Bateson, G. (2000) *Steps to an Ecology of Mind* (Chicago, University of Chicago Press).

Beck, U. (1992) *Risk Society: Towards a New Modernity* (London, Sage).

Berlin, I. (2000a) *The Roots of Romanticism* (London, Pimlico).

Berlin, I. (2000b) *Three Critics of the Enlightenment* (London, Pimlico).

Bernal, D. (1969) *Science in History*, Vol. 3 (Harmondsworth, Pelican Books).

Bernstein, B. (1965) A socio-linguistic approach to social learning, in: J. Gould (ed.) *Social Science Survey* (Harmondsworth, Penguin Books).

Biesta, G. (2006) *Beyond Learning* (Boulder, CO, Paradigm Publishers).

Blenkinsop, S., Affifi, R., Piersol, L. and Sitka-Sage, M. (2017) Shut-up and listen: Implications and possibilities of Albert Memmi's characteristics of colonization upon the "Natural World", *Studies in Philosophy and Education*, 36(3), pp. 349–365.

Bluhdorn, I. (2002) Unsustainability as a frame of mind – And how we disguise it: The silent counter revolution and the politics of simulation, *The Trumpeter*, 18(1), pp. 59–69.

Blühdorn, I. (2010) Political sociology and the cultural framing of environmental discourse: Depoliticisation, repoliticisation and the governance of unsustainability. Unpublished paper presented to AHRC network The Cultural Framing of Environmental Discourse, Workshop I, 2–3 December 2010, Bath, UK.

Bonnett, M. (1978) Authenticity and education, *Journal of Philosophy of Education*, 12, pp. 51–61.

Bonnett, M. (1997) Environmental education and beyond, *Journal of Philosophy of Education*, 31(2), pp. 249–266.

Bonnett, M. (2004) *Retrieving Nature: Education for a Post-Humanist Age* (Oxford, Wiley-Blackwell).

Bonnett, M. (2007) Environmental education and the issue of nature, *Journal of Curriculum Studies*, 39(6), pp. 707–721.

Bonnett, M. (2009a) Education and selfhood: A phenomenological investigation, Special Issue *Journal of Philosophy of Education: What do Philosophers of Education Do?* 43(3), pp. 357–370.

Bonnett, M. (2009b) Schools as places of unselving: An educational pathology? In: G. Dall'Alba (ed.) *Exploring Education Through Phenomenology. Diverse Approaches* (Oxford, Wiley-Blackwell).

Bonnett, M. (2009c) Systemic wisdom, the 'selving' of nature, and knowledge transformation: Education for the 'greater whole', *Studies in Philosophy and Education*, 28, pp. 39–49.

Bonnett, M. (2010) Getting and spending, we lay waste our powers: Environmental education and the culture of the school, *FORUM*, 52(1), pp. 85–90.

Bonnett, M. (2012) Environmental concern, moral education, and our place in nature, *Journal of Moral Education Special Issue: Moral Education and Environmental Concern*, 41(3), pp. 285–300.

Bonnett, M. (2013a) Sustainable development, environmental education, and the significance of being in place, *Curriculum Journal*, 24(2), pp. 250–271.

Bonnett, M. (2013b) Normalizing catastrophe: Sustainability and scientism, *Environmental Education Research*, 19(2), pp. 187–197.

Bonnett, M. (2013c) Self, environment, and education: Normative arisings, in: M. Brody, J. Dillon and R. Stevenson (eds.) *International Handbook of Research on Environmental Education* (New York, Routledge).

Bonnett, M. (2015a) Sustainability, the metaphysics of mastery, and transcendent nature, in: H. Kopnina and E. Shoreman-Ouimet (eds.) *Sustainability: Key Issues* (London, Routledge).

Bonnett, M. (2015b) The powers that be: Environmental education and the transcendent, *Policy Futures in Education*, 13(1), pp. 42–56.

Bonnett, M. (2015c) Transcendental nature and the character of truth and knowledge in education, in: P. Kemp and S. Frolund (eds.) *Nature in Education* (Zurich, LIT-Verlag).

Bonnett, M. (2017a) Sustainability and human being: Towards the hidden Centre of authentic education, in: B. Jickling and S. Sterling (eds.) *Post-Sustainability and Environmental Education* (Cham, Switzerland, Palgrave Macmillan).

Bonnett, M. (2017b) Environmental consciousness, sustainability, and the character of philosophy of education, *Studies in Philosophy and Education*, 36(3), pp. 333–347.

Bowers, C. (2002) Toward an eco-justice pedagogy, *Environmental Education Research*, 8(1), pp. 21–34.

Bowers, C. (2011) *Perspectives on the Ideas of Gregory Bateson, Ecological Intelligence, and Educational Reforms* (Eugene, OR, Eco-justice Press).

Brentano, F. (1995) *Psychology from an Empirical Standpoint* (London, Routledge).

Brundtland Commission (1987) *Our common future* (Milton Keynes, Open University Press).

Butler, J. (1997) *Excitable Speech. A Politics of the Performative* (New York, Routledge).

Callicot, J. (1986) On the intrinsic value of nonhuman species, in: B. Norton (ed.) *The Preservation of Species: The Value of Biological Diversity* (Princeton, NJ, Princeton University Press), pp. 138–172.

Callicot, J. (1995) Animal liberation: A triangular affair, in: R. Elliot (ed.) *Environmental Ethics* (Oxford, Oxford University Press).
Capra, F. (1982) *The Turning Point* (New York, Simon & Schuster).
Carson, R. (1962) *Silent Spring* (Boston, Houghton Mifflin).
Chang, M. (2020) Encounters with suchness: Contemplative wonder in environmental education, *Environmental Education Research*, 26(1), pp. 1–13.
Chinnery, A. (2018) Emmanuel Levinas, autonomy, and education, in: P. Smeyers (ed.) *International Handbook of Philosophy of Education, Part 1* (Cham, Switzerland, Springer).
Crist, E. (2008) Against the social construction of nature and wilderness, in: M. Nelson and J. Callicott (eds.) *The Wilderness Debate Rages on Volume II* (Georgia, University of Georgia Press).
Crist, E. (2012) Abundant earth and the population question, in: P. Cafaro and E. Crist (eds.) *Life on the Brink: Environmentalists Confront Overpopulation* (Georgia, University of Georgia Press).
Crutzen, P. and Stoermer, E. (2000) The 'Anthropocene', *Global Change Newsletter*, 41, pp. 17–18.
Daily, G., Ehrlich, P. and Ehrlich, A. (1994) Optimum human population size, *Population and Environment*, 15(6), pp. 469–475.
Darwin, C. (1872) *The Origin of Species* (Garden City, NY, Doubleday).
Elliott, J. (1999) Sustainable society and environmental education: Future perspectives and demands for the education system, *Cambridge Journal of Education*, 29(3), 325–340.
Farrelly, M. (2019) The significance of myth for environmental education, *Journal of Philosophy of Education*, 53(1), pp. 127–144.
Ford, D. and Blenkinsop, S. (2018) Learning to speak Franklin: Nature as co-teacher, *Journal of Outdoor and Environmental Education*, 21(3), pp. 307–318.
Foreman, D. (1993) Putting the earth first, in: S. Armstrong and R. Botzler (eds.) *Environmental Ethics* (New York, McGraw-Hill).
Garrard, G. (1998) The romantics' view of nature, in: D. Cooper and J. Palmer (eds.) *Spirit of the Environment* (London, Routledge).
Giddens, A. (1994) *Beyond Left and Right* (Cambridge, Polity Press).
Glacken, C. (1967) *Traces on the Rhodian Shore* (Berkeley, University of California Press).
Gonzalez-Gaudiano, E. (2006) Environmental education: A field in tension or transition? *Environmental Education Research*, 12(3–4), pp. 291–300.
Gould, S. (1990) The golden rule – A proper scale for our environmental crisis, *Natural History* (September), pp. 24–30.
Grange, J. (1997) *Nature. An Environmental Cosmology* (Albany, State University of New York).
Griffiths, M. and Murray, R. (2017) Love and justice in learning for sustainability, *Ethics and Education*, 12(1), pp. 39–50.
Heidegger, M. (1962) *Being and Time*, trans. J. Macquarrie and E. Robinson (Oxford, Wiley-Blackwell).
Heidegger, M. (1975) *Poetry, Language, Thought*, trans. A. Hofstadter (New York, Harper & Row).
Heidegger, M. (1977) The question concerning technology, in: W. Lovitt (trans.) *The Question Concerning Technology and Other Essays* (New York, Harper & Row).
Heidegger, M. (1978) The essence of truth, in: D. Krell (ed.) *Martin Heidegger Basic Writings* (London, Routledge and Kegan Paul).
Heise, U. (2008) *Sense of Place and Sense of Planet* (Oxford, Oxford University Press).

Hepburn, R. (2004) Landscape and metaphysical imagination, in: A. Carlson and A. Berleant (eds.) *The Aesthetics of Natural Environments* (Peterborough, ON, Broadview Press).

Husserl, E. (2001) *Logical Investigations Volume II* (London, Routledge).

IUCN, UNEP, WWF, FAO, & UNESCO (1980) *World Conservation Strategy* (Zurich, IUCN-UNEP-WWF). Retrieved from http://data.iucn.org/dbtw-wpd/edocs/WCS004.pdf.

Jickling, B. (2013) Normalising catastrophe: An educational response, *Environmental Education Research*, 19(2), pp. 161–176.

Johannesdottir, G. and Thorgeirsdottir, S. (2015) Understanding our place in the natural world, coming to our senses through embodied experiences of ecophilosophical and posthumanist art, in: P. Kemp and S. Frolund (eds.) *Nature in Education* (Zurich, Lit-Verlag).

Johnson, K. (2008) Natural history as stamp collecting: A brief history, *Archives of Natural History*, 34(2), pp. 244–258.

Kant, E. (1933) *Immanuel Kant's Critique of Pure Reason* (London, Palgrave Macmillan).

Kemp, P. (2017) The animal – My partner, in: P. Kemp and S. Frolund (eds.) *Nature in Education* (Zurich, Lit-Verlag).

Kenklies, K. (2020) Dogen's time and the flow of otiosity – Exiting the educational rat race, *Journal of Philosophy of Education*. https://doi.org/10.1111/1467-9752.12410.

Kopnina, H. (2012) Education for sustainable development (ESD): The turn away from 'environment' in environmental education? *Environmental Education Research*, 18(5), pp. 699–717.

Kopnina, H. (2016) Half the earth for people (or more)? Addressing ethical questions in conservation, *Biological Conservation*, 203, pp. 176–185.

Kopnina, H. and Washington, H. (2016) Discussing why population growth is still ignored or denied, *Chinese Journal of Population Resources and Environment*, pp. 1–11.

Kopnina, H., Washington, H., Gray, J. and Taylor, B. (2018) The 'future of conservation' debate: Defending ecocentrism and the nature needs half movement, *Biological Conservation*, 217, pp. 140–148.

Leopald, A. (1993) The land ethic, in: S. Armstrong and R. Botzler (eds.) *Environmental Ethics. Divergence and Convergence* (New York, McGraw-Hill).

Levinas, E. (1981) *Otherwise than Being or Beyond Essence* (The Hague, Martinus Nijhoff).

Louv, R. (2010) *Last Child in the Woods* (London: Atlantic Books).

Louv, R. (2012) *The Nature Principle* (Chapel Hill, NC, Algonquin Books).

Lovelock, J. (2009) *The Vanishing Face of Gaia* (New York, Basic Books).

Lyotard, J.-F. (1984) *The Postmodern Condition: A Report on Knowledge* (Manchester, Manchester University Press).

Magrini, J. (2019) *Ethical Response to Nature's Call: Reticent Imperatives* (London, Routledge).

Marshall, P. (1995) *Nature's Web* (London, Cassell).

Matthews, F. (1994) *The Ecological Self* (London, Routledge).

McDaniel, J. (1986) A feeling for the organism: Christian spirituality as openness to fellow creatures, *Environmental Ethics*, 8, pp. 33–46.

McDowell, J. (1996) *Mind and World* (Cambridge, MA, Harvard University Press).

McKibben, W. (1989) *The End of Nature* (New York, Random House).

Merleau-Ponty, M. (1962) *The Phenomenology of Perception* (London, Routledge and Kegan Paul).

Moore, G. E. (1953) *Some Main Problems of Philosophy* (London: Allen and Unwin).

Morse, M., Jickling, B. and Morse, P. (2018) Views from a pinhole: Experiments in wild pedagogy on the Franklin River, *Journal of Outdoor and Environmental Education*, 21(3), pp. 255–275.

Morse, M., Jickling, B. and Quay, B. (2018) Rethinking relationships through education: Wild pedagogies in practice, *Journal of Outdoor and Environmental Education*, 21(3), pp. 241–254.

Murdy, W. (1975) Anthropocentrism: A modern version, *Science*, 187, pp. 1168–1172.

Noss, R. and Cooperrider, A. (1994) *Saving Nature's Legacy: Protecting and Restoring Biodiversity* (Washington, DC, Island Press).

Oakeshott, M. (1972) Education: The engagement and its frustration, in: R. Dearden, P. Hirst and R. Peters (eds.) *Education and the Development of Reason* (London, Routledge and Kegan Paul).

Olvitt, L. (2017) Education in the Anthropocene: Ethico-moral dimensions and critical realist openings, *Journal of Moral Education*, 46(4), pp. 396–409.

Orr, D. (1994) *Earth in Mind: On Education, the Environment and the Human Prospect* (Washington, DC, Island Press).

Østergaard, E. (2019) Music and sustainability education – A contradiction? *Acta Didactica Norge*, 13(2), Art. 2, 20. https://doi.org/10.5617/adno.6452.

Palmer, J. (1999) Research matters: A call for the application of empirical evidence to the task of improving the quality and impact of environmental education, *Cambridge Journal of Education*, 29(3), pp. 379–395.

Payne, P. G. (2006) The technics of environmental education, *Environmental Education Research*, 12(3–4), pp. 487–502.

Payne, P. G. and Wattchow, B. (2009) Phenomenological deconstruction, slow pedagogy and the corporeal turn in wild environmental/outdoor education, *Canadian Journal of Environmental Education*, 14, pp. 15–32.

Peters, R. (1959) *Authority, Responsibility and Education* (London, George Allen and Unwin).

Posch, P. (1999) The ecologisation of schools and its implications for educational policy, *Cambridge Journal of Education*, 29(3), pp. 341–348.

Postma, D. (2006) *Why Care for Nature? In Search of an Ethical Framework for Environmental Responsibility and Education* (Dordrecht, Springer).

Price, H. H. (1932) *Perception* (London, Methuen).

Robottom, I. (2005) Critical environmental education research: Re-engaging the debate, *Canadian Journal of Environmental Education*, 10, pp. 62–78.

Rolston III, H. (1997a) Nature for real: Is nature a social construct? In: T. Chapell (ed.) *The Philosophy of the Environment* (Edinburgh, Edinburgh University Press).

Rolston III, H. (1997b) Ethics on the home planet, in: A. Watson (ed.) *An Invitation to Environmental Philosophy* (New York, Oxford university Press).

Rorty, R. (1980) *Philosophy and the Mirror of Nature* (Oxford, Blackwell).

Russell, B. (1959) *The Problems of Philosophy London* (Oxford, Oxford University Press).

Sahlins, M. (1977) *The Use and Abuse of Biology* (London, Tavistock Publishers).

Scheler, M. (1980) *The Collected Works of Max Scheler*, Vol VIII (Bern, Francke Ver-lag).

Schinkel, A. (2017) The educational importance of deep wonder, *Journal of Philosophy of Education*, 51(2), pp. 538–553.

Scrimshaw, P. (1989) Educational computing: What can philosophy of education contribute? *Journal of Philosophy of Education*, 23(1), pp. 103–111.

Seigel, J. (2005) *The Idea of the Self* (Cambridge, Cambridge University Press).

Sharp, H. (1988) *The Transformation of Bigfoot: Madness, Power and Belief among the Chipewyan* (Washington, DC: Smithsonian Institution Press).

Shuffleton, A. (2017) Jean-Jacques Rousseau, the mechanical clock and children's time, *Journal of Philosophy of Education*, 51(4), pp. 837–849.

Singer, P. (1993) *Practical Ethics*, Second Edition (Cambridge, Cambridge University Press).

Skulason, P. (2015) The wildness of nature. Its significance for our understanding of the world, in: P. Kemp and S. Frolund (eds.) *Nature in Education* (Zurich, Lit Verlag).

Solomon, R. (1980) *History and Human Nature* (Brighton, UK: Harvester Press).

Stables, A. (2009) The unnatural nature of nature and nurture: Questioning the romantic heritage, *Studies in Philosophy and Education*, 28(3), pp. 3–14.

Sterling, S. (2017) Assuming the future: Repurposing education in a volatile age, in: B. Jickling and S. Sterling (eds.) *Post-Sustainability and Environmental Education* (Cham, Switzerland, Palgrave Macmillan).

Stevenson, R. (2006) Tensions and transitions in policy discourse: Recontextualizing a decontextualized EE/ESD debate, *Environmental Education Research*, 12(3–4), pp. 277–290.

Stickney, J. (forthcoming) Seeing trees: Investigating the poetics of place-based, aesthetic environmental education with Heidegger and Wittgenstein, *Journal of Philosophy of Education*.

Stone, C. D. (1974) *Should Trees Have Standing? Towards Legal Rights for Natural Objects* (Los Altos, CA, William Kaufman).

Taylor, P. (1986) *Respect for Nature: A Theory of Environmental Ethics* (Princeton, NJ, Princeton University Press).

UN (2017) *World Population Prospect: The 2017 Revision* (New York, United Nations).

United Nations Conference on Environment and Development (UNCED) (1992) *Agenda 21* (New York: UNCED).

United Nations Environment Programme (UNEP) (2012) *GEO 5 Report: Environment for the Future We Want* (Nairobi, UNEP).

United Nations Environment Programme (UNEP) (2019) *GEO 6 Report: Healthy Planet, Healthy People* (Cambridge, Cambridge University Press).

United Nations Educational, Scientific and Cultural Organization (UNESCO) (2005) *Education for Sustainable Development* (Paris, UNESCO).

Wark, M. (1994) Third nature, *Cultural Studies*, 8, pp. 115–132.

Washington, H. (2018) Education for wonder, *Education Sciences*, 8(3), 125.

Waters, C., Zalasiewicz, J., Summerhayes, C., Barnosky, A., *et al.* (2016) Review summary: The Anthropocene is functionally and stratigraphically distinct from the Holocene, *Science*, 351, p. 137.

Weil, S. (2005) *Simone Weil: An Anthology* (London, Penguin Books).

Wellington, B. (2011) The myth of mountaintop removal mining, *The Guardian*, 19 August.

Whitehead, A. (1925) *Science and the Modern World* (New York, Palgrave Macmillan).

Wilson, O. (2016) *Half-Earth. Our Planet's Fight for Life* (London, Liveright Publishing).

Wood, D. (2003) What is Eco-Phenomenology? In: C. Brown and T. Toadvine (eds.) *Eco-Phenomenology: Back to the Earth Itself* (Albany, SUNY Press), pp. 211–234.

WWF (2014) *Living Planet Index* (Gland, Switzerland, WWF).

INDEX

For Product Safety Concerns and Information please contact our EU representative GPSR@taylorandfrancis.com
Taylor & Francis Verlag GmbH, Kaufingerstraße 24, 80331 München, Germany

www.ingramcontent.com/pod-product-compliance
Ingram Content Group UK Ltd.
Pitfield, Milton Keynes, MK11 3LW, UK
UKHW020137200926
475702UK00005B/797
9780367373436